3rd International Conference on Architecture and Civil Engineering (ICACE 2019)

AA007584

IOP Conference Series: Materials Science and Engineering
Volume 636

Kuala Lumpur, Malaysia
26 - 27 June 2019

Editors:

Sattar S. Emamian
Meor Mohammad Fared Meor Razali
Mokhtar Awang

ISBN: 978-1-5108-9838-7
ISSN: 1757-8981

Printed from e-media with permission by:

Curran Associates, Inc.
57 Morehouse Lane
Red Hook, NY 12571

Some format issues inherent in the e-media version may also appear in this print version.

This work is licensed under a Creative Commons Attribution 3.0 International Licence. Licence details: http://creativecommons.org/licenses/by/3.0/.

No changes have been made to the content of these proceedings. There may be changes to pagination and minor adjustments for aesthetics.

Printed with permission by Curran Associates, Inc. (2020)

For permission requests, please contact the Institute of Physics at the address below.

Institute of Physics
Dirac House, Temple Back
Bristol BS1 6BE UK

Phone: 44 1 17 929 7481
Fax: 44 1 17 920 0979

techtracking@iop.org

Additional copies of this publication are available from:

Curran Associates, Inc.
57 Morehouse Lane
Red Hook, NY 12571 USA
Phone: 845-758-0400
Fax: 845-758-2633
Email: curran@proceedings.com
Web: www.proceedings.com

Table of contents

Volume 636

3rd International Conference on Architecture and Civil Engineering (ICACE 2019)

26–27 June 2019, Kuala Lumpur, Malaysia

Accepted papers received: 30 August 2019
Published online: 15 October 2019

Preface

Papers

IOP Conf. Series: Materials Science and Engineering **636** (2019) 011001 doi:10.1088/1757-899X/636/1/011001

Preface:

This volume presents a compilation of research works covering the fields of Architecture and Civil Engineering. All manuscripts in this volume were presented during the 3rd International Conference on Architecture and Civil Engineering 2019 (ICACE 2019) which was held at Seri Pacific Hotel, Kuala Lumpur, Malaysia on 26-27 June 2019. The editor(s) of the proceeding would like to express the utmost gratitude and thanks to all reviewers in the technical team for making this volume a success.

Editors:

Dr. Sattar S. Emamian

Ar. Meor Mohammad Fared Meor Razali

Ir. Dr. Mokhtar Awang

Content from this work may be used under the terms of the Creative Commons Attribution 3.0 licence. Any further distribution of this work must maintain attribution to the author(s) and the title of the work, journal citation and DOI.
Published under licence by IOP Publishing Ltd

Content from this work may be used under the terms of the Creative Commons Attribution 3.0 licence. Any further distribution of this work must maintain attribution to the author(s) and the title of the work, journal citation and DOI.
Published under licence by IOP Publishing Ltd

IOP Conf. Series: Materials Science and Engineering **636** (2019) 011003 doi:10.1088/1757-899X/636/1/011003

List of Papers for IOP			
No	Paper ID	Corresponding Author	Title
1	191604	Shofy Nur Fajri	Lineament Analysis Of Digital Elevation Model To Identification Of Geological Structure In Northern Manna Sub-Basin, Bengkulu
2	195644	Cecilia May Villanueva	Transformations of Urban Space in Community and Pedestrian Wellbeing
3	1910826	Wong Leong Yee	Application of parametric components in architectural planning
4	1912043	Xia Qing	Invasion of China's Traditional Architecture: A Review
5	1912328	Khalida Muda	Long-Range Transport and Local Emission of Atmospheric PM 2.5 in Southern Region of Peninsular Malaysia
6	1912629	Alaa Abdalla Saeid Ali	Theoretical Performance of a Multi-Cyclone Fine Particulate Emission Control System
7	1912730	Nur Syamimi Zaidi	Effectiveness of Local Waste Materials as Organic-Based Coagulant in Treating Water
8	1913034	Shuhairy Norhisham	Service Frequency And Service Hours Evaluation For Bus Service In West Klang Valley
9	1913537	Alaa Abdalla Saeid Ali	Light Pollution and Health: Case Study of the Lighting Fixtures Applied on Penang Town Hall in George Town, Penang Island
10	1913648	Mari TS	Architecture Graduate Work Readiness: The Gap Between Learning and Employability
11	1914271	Siti Balkish Roslan	Layering of History in Malaysian Modern Architecture through Analytical Diagramming
12	1914538	Veronica Ng	Chinese Cemetery as Heterotopia: A Case of Kwong Tong Chinese Cemetery (KTCC)
13	1914840	Mohd Jaki Mamat	Integrated space in the traditional Pondok school education system - A case study in Pattani, Thailand.
14	1915541	Noran Abdulhameed	Adaptive Reuse Approaches Of Shophouses At Cannon Street In George Town, Penang
15	1915945	Sujatavani Gunasagaran	Adoption of Social Media by Architecture Students in Fostering Community Service Initiative using Technology Acceptance Model
16	1916169	Sucharita Srirangam	Typologies of Design Thinking: The learner-led perspectives
17	1916850	Molood Seifi	Creating Secured Residential Places: Conflicting Design Elements of Natural Surveillance, Access Control and Territoriality
18	1917353	Nurhaya Baniyamin	Design Strategies In Science Centre
19	1917864	Woon Yoke Bee	Community Rainwater Harvesting Financial Payback Analyses - Case Study In Malaysia
20	1918660	Sheikh Ilmi, Habizah	Employee Turnover of Quantity Surveying Firms in Malaysia
21	1919266	Doddy Friestya Asharsinyo	Degree Level of Publicness Through Meaning of Public Sphere In Bandung City, West Java, Indonesia
22	1919977	Mohammad Arif Rohman	Understanding Social Sustainability Criteria for Public School Building

Content from this work may be used under the terms of the Creative Commons Attribution 3.0 licence. Any further distribution of this work must maintain attribution to the author(s) and the title of the work, journal citation and DOI.
Published under licence by IOP Publishing Ltd

Peer review statement

All papers published in this volume of *IOP Conference Series: Materials Science and Engineering* have been peer reviewed through processes administered by the proceedings Editors. Reviews were conducted by expert referees to the professional and scientific standards expected of a proceedings journal published by IOP Publishing.

Content from this work may be used under the terms of the Creative Commons Attribution 3.0 licence. Any further distribution of this work must maintain attribution to the author(s) and the title of the work, journal citation and DOI.
Published under licence by IOP Publishing Ltd

IOP Conf. Series: Materials Science and Engineering **636** (2019) 012001 doi:10.1088/1757-899X/636/1/012001

Lineament analysis of digital elevation model to identification of geological structure in Northern Manna Sub-Basin, Bengkulu

SN Fajri[1] E Surtiyono[1] and S Nalendra[1]
[1] Geological Engineering Study Program, University of Sriwijaya, Palembang, Indonesia
E-mail: shofynurfajri@gmail.com

Abstract. Digital Elevation Model (DEM) is a representation of the actual landform based on elevation. DEM is part of the technological sophistication that is very helpful in planning, mapping, and surveying. This study applies DEM, which is constructed from DEMNAS, focusing on DEM optimization for geological structures identification.The applied method is ellipsoid stress model analysis approach to determine the type of structure. The scope of this study is located in North Manna Sub-Basin, tectonically is a basin that formed on Paleogen by an extensional force that forms half-graben along the Bengkulu Shelf. Interpretation of DEM in the study area showed the existence of 10 lineament patterns of NE-SW and five NNE-SSW patterns, controlled by normal fault, nine lineament patterns of NW-SE and six WNW-ESE patterns, controlled by a thrust fault and fold. Field observation using direct measurement method obtained by four faults with NE-SW and NNE-SSW pattern, one fault and four folds with NW-SE pattern,three faults and two folds with WNW-ESE trending pattern. This indicates that the accuracy level of DEM in the study area reaches 90%. The geological structure that is still active, certainly requires special attention in building construction.

1. Introduction

Technological advances and sophistication, provide many conveniences in a variety of areas, one of which is geological surveying and mapping. DEM is one of the sophistications that is very useful in that field. According to Burrough in [7], DEM is a representation of the actual surface of the earth with a certain degree of accuracy. The level of DEM's precision to the actual situation depends on the resolution and the source of the DEM itself [3], [7]. DEM can be used in making maps of slopes, gradients, other aspects, and various interests such as geological analysis [2].

Identifying the geological structure using DEM has become something quite actual, because data that can be analyzed is a lineament illustrated by its three dimensional representation. DEM data provides vertical image reflection of the presence of faults and folds below the surface [4]. This study uses DEM from SRTM to test its accuracy in the identification of geological structures. Lineament analysis using DEM is the best way to get an initial illustration of the geological conditions of a place [16]. The use of DEM from SRTM for similar analysis is also done by [20], and yields data that is

Content from this work may be used under the terms of the Creative Commons Attribution 3.0 licence. Any further distribution of this work must maintain attribution to the author(s) and the title of the work, journal citation and DOI.
Published under licence by IOP Publishing Ltd

IOP Conf. Series: Materials Science and Engineering **636** (2019) 012001 doi:10.1088/1757-899X/636/1/012001

good enough to be applied to other geological aspects, such as geodynamics, mineral exploration, and hydrogeology. The SRTM level accuracy test has also been done by [8], [14], and [6].

The study area has an area of 81 km^2, administratively located in South Bengkulu District, Bengkulu Province, Indonesia (Fig 1). Geologically, it is the northern part of the Manna Sub-Basin which in [23], it is referred to as Kedurang Graben. According to [5], the geological structure in this area is highly developed with strong influence from Sumatran Fault System (SFS).

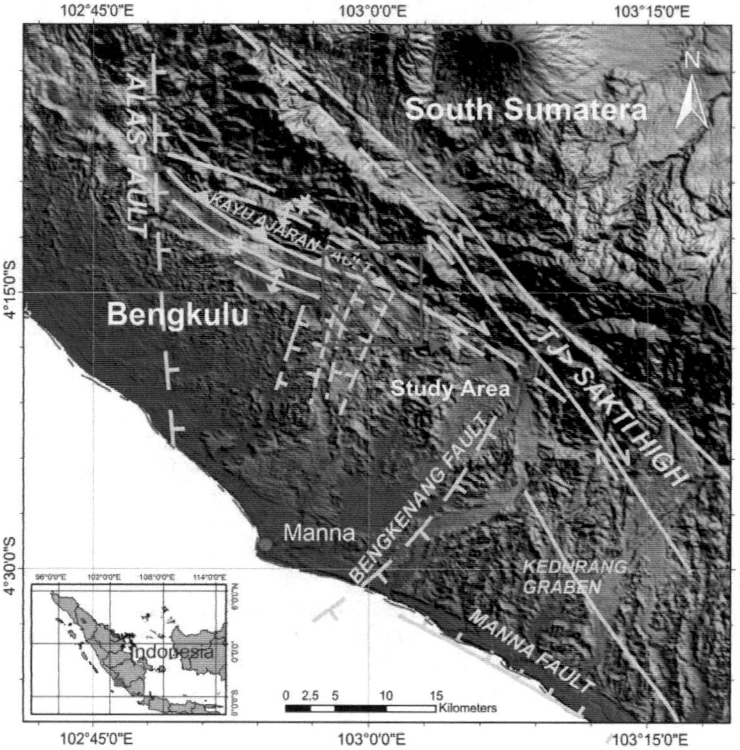

Fig. 1 The SRTM image illustrates that the northern morphology of the study area is a ridge with NW-SE direction (modification from [23]).

2. An overview of geology

The subduction system of Sumatra has undergone three times of change of direction, namely the NW-SE direction of the Late Jura-Late Cretaceous with the compression phase, N-S direction in Late Cretaceous-Early Tertiary with the tensional phase, and the NE-SW direction at Middle Miocene-Recent by compression phase [19]. At Tertiary-Recent, subduction forms an angle of N25^{O}E in the southern part of Sumatra and forms an angle N31^{O}E on the northern part of Sumatra, this forming a major fault on the island of Sumatra with the movement of strike-slip fault namely Semangko Fault [18], [22], and [9]. The Bengkulu Basin itself began to develop in the Early Tertiary [9], [11]. This sedimentary basin then decreased slowly in the Early Miocene [12], [15]. At the Middle Miocene the removal of the Barisan Mountains lane triggers magmatic activity around it [13]. The process of descending the basin ends at the Late Miocene-Pliocene [23].

Tectonically, study area is dominated by the movement of the Sumatran Fault System (SFS). According to [21], the SFS along the Bengkulu Basin is divided into several segments, from the south, namely Kumering Segment, Manna, Musi, and Ketaun Segment. Study area, based on [21], is in the southern part of the Manna Segment (Fig. 2). The Manna Segment has a length of 85 km, with morphological dominance of mountain ranges in the eastern fault zone [17]. According to [17], this segment is still active with evidence of an earthquake in 1893.

The stratigraphic sequence of study areas, from oldest to youngest, are Seblat Formation, Lemau Formation, and Simpangaur Formation [1]. Seblat Formations in the form of claystone, carbonaceous claystone, and sandstone with limestone insertions are deposited in conformity over the Hulusimpang Formation in the Early-Middle Miocene [1]. Then the Lemau Formation, which is composed of claystone, carbonaceous claystone, coal, sandstones, conglomerates, and breccias from the precipitated alignment above the Seblat Formation on Middle-End Miocene [23]. After that, it is precipitously aligned with the Simpangaur Formation with a conglomeratic sandstone, sandstones, claystone, and molluscas sandstone, and tufsandstone at the Late Miocene-Pliocene [1].

The geological structure of the study area, according to [23], is divided into two general directions, namely the NW-SE and NE-SW directions. Fault with dextral movement (Tanjung Sakti Fault, Ketaun Fault, and Manna Fault) is the controller of Pagarjati Sub-basin (Pagarjati Graben) in NW and Manna or Kedurang Sub-basin (Kedurang Graben) as study area in SE part of Bengkulu Basin [23]. The geological structure of the study area is regionally controlled by the Tanjung Sakti Fault, which is part of the Semangko Fault lane [23].

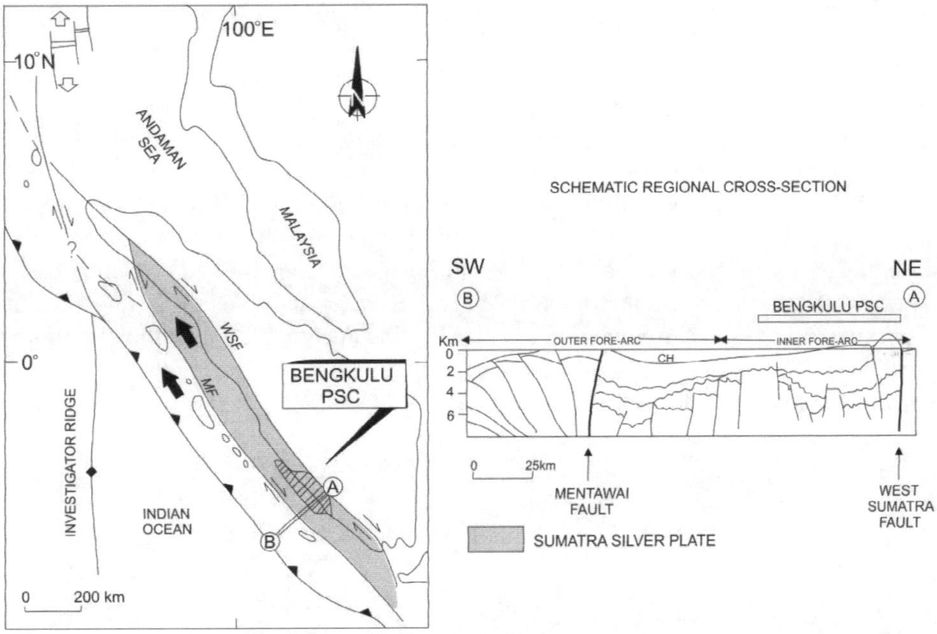

Fig. 2 Tectonic of Bengkulu Basin [9].

3. Methodology

Process of DEM maps making is using a global mapper and arcGIS software. The data used for this study is an elevation model with accuracy of 0,27-arcsecond (8,25 meters) from the web of DEMNAS, the Indonesian Geographic Information Agency. The procedure begins with opening elevation data in global mapper, then creating contour of the study area by using contour generate menu. The contours created must be in the projected UTM coordinates. Once the contour is finished, it is saved into a shapefile, next, opens the contour file with the arcgis program. To be a DEM, contour data goes through two processes, namely the conversion of a file into a raster form, and then into a TIN. If the process succeeds, it will form the desired DEM.

The lineament pattern is done by delineating the form of lineament reflected from the appearance of three dimensional reliefs of DEM. After delineation, azimuth measurements, length of alignment, and number of lineaments are found in each zone of coordinate grid. The calculation of the number of lineaments is based on azimuth orientation. It is intended to know the intensity of the lineament in each zone. Comparison of lineament intensity is visualized into several forms like contours, graphs, and rose diagrams. Contours are created using the help of surfer programs, graphics using StatPlus, and rose diagrams using the dips program.

An analysis of the geological structure of the lineament was performed using an ellipsoid stress model [10]. The major axial direction used is NE-SW or Miocene-Recent decryption determined from previous studies [19], [9], [18], [22]. It is also assessed from the rock formations that make up the study area which is the Miocene Formation, so it is concluded that the structure of this region certainly has the same age or is younger than the age of formation.

The accuracy of alignment analysis derived from the percentage comparison of DEM's geological structure analysis results with the availability of geological structures in the field. Parameters are assessed in the form of azimuth similarity, number of straightness, and structure type accuracy.

4. Result and discussion

The DEM maps created in this study visualize a three-dimensional relief form with a shadow derived from a solar lighting scenario with a 30° angle at azimuth N315^{o}E (Fig. 3). Then after the delineation, obtained 36 directions of lineament (Fig. 4). In the previous study, the lineament was divided into several groups, namely N-S, E-W, NE-SW, NW-SE, NNE-SSW, ENE-WSW, WNW-ESE, and NNW-SSE [21]. In this study, the alignment encountered has N-S, E-W, NE-SW, NW-SE, NNE-SSW, and WNW-ESE orientations (Table 1). The lineament dominance occurs at the NE-SW orientation with ten straightnesses as results indicate, and the NW-SE orientation with nine straightnesses, while the least orientation is N-S and E-W orientations with the amount of three straightnesses in each. The most dominant percentage of lineament is 28% and the minimum is 8%.

Table 1 Length and percentage of lineament

Length (km)	N-S	E-W	NE-SW	NW-SE	NNE-SSW	WNW-ESE
0-0,8	-	-	2	-	-	1
0,8-1,6	3	3	7	7	5	3
>1,6	-	-	1	2	-	2
Sum	3	3	10	9	5	6
%	8	8	28	25	14	17

As observed from the rose diagram, lineament in the study area also shows the NE-SW and NW-SE patterns (Fig. 5a). The alignment with the N-S and NE-SW directions is dominant in the Northeast study location. In the North, the lineaments with dominancy are the NW-SE and NNE-SSW patterns, while the Southern part is dominated by the E-W. The northwestern of the study area is dominated by the WNW-ESE lineament (Fig. 6a).

A fairly long range of alignment (> 1.6 km), is predominantly stretched from the NE-SW and SW-SE of study area (Fig. 6b). The pattern has NE-SW, NW-SE, and WNW-ESE orientations. Comparison between the length of straightness and azimuth orientation indicates that the study area is dominated by lineament in the direction of 0-90° and 270-360°, and has a length between 0.5-2.7 km (Fig. 5b).

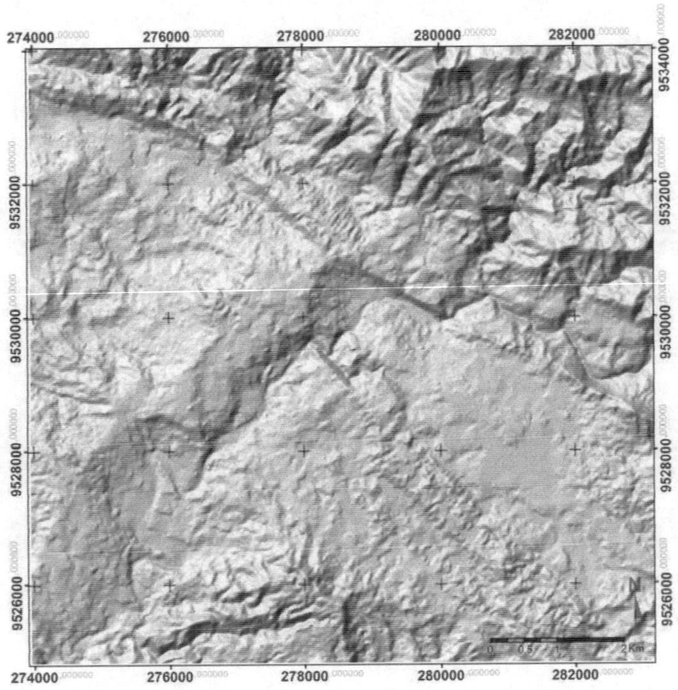

Fig. 3 Appearance of DEM relief in the study area.

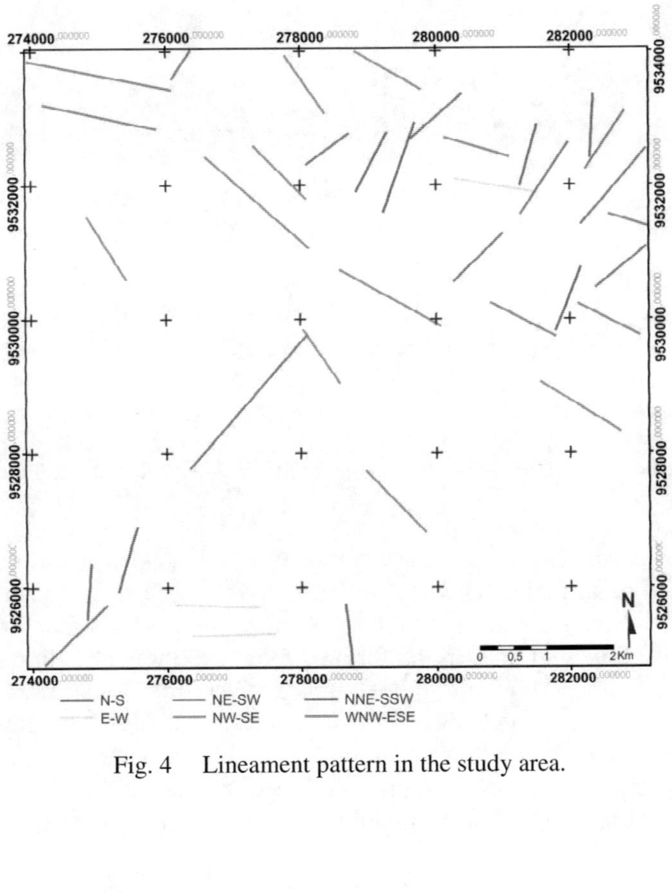

Fig. 4 Lineament pattern in the study area.

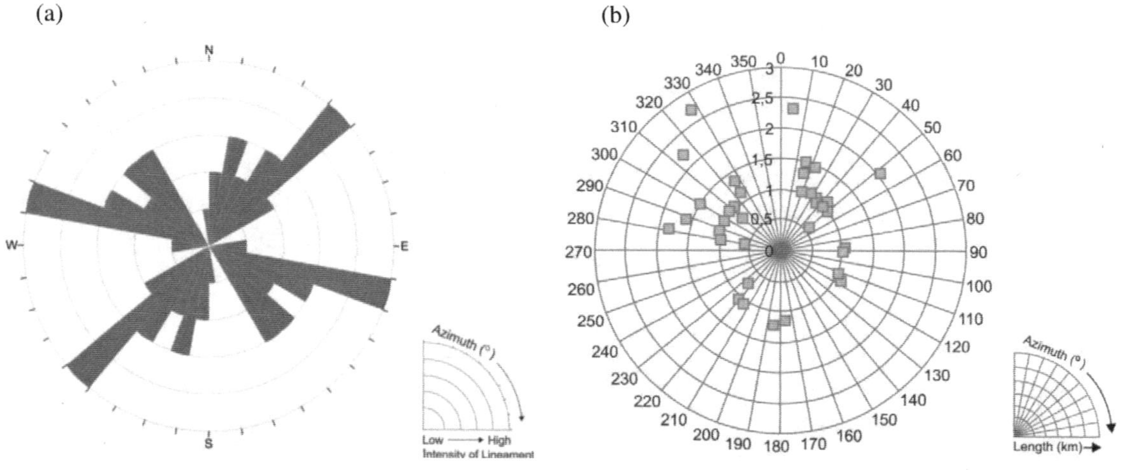

Fig. 5 (a) Lineament orientation in form rose diagram, (b) Comparison of azimuth value and length of alignment.

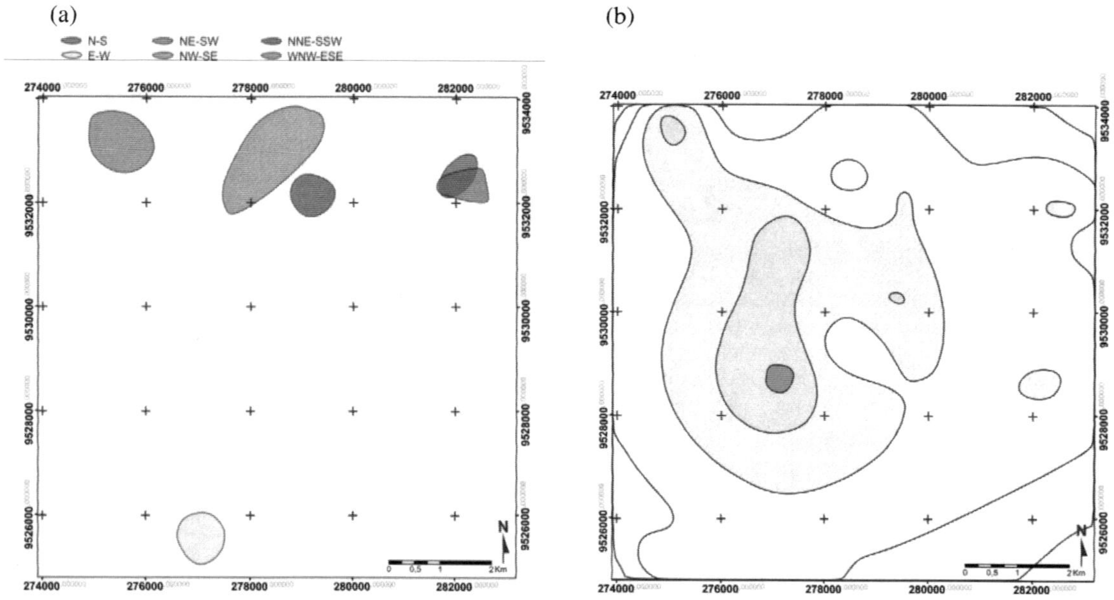

Fig. 6 (a) Isodominance of lineament based on intensity, (b) Isolength lineament (orange shows high value).

Results from the analysis with ellipsoid stress model [10], lineament with NE-SW and NNE-SSW directions, 100% are normal faults. This is directly proportional to the structure found in the field, where the orientation is entirely a normal fault. Lineament with orientations of NW-SE and WNW-ESE, has 50% of thrust fault and 50% in the form of folds, but the reality in the field shows 40% of the alignment is thrust fault and 60% are folds. In this analysis, there is no structure with orientations of N-S and E-W. Likewise, with the reality in the field, it is possible that the lineament N-S is part of the normal fault, while the E-W is a thrust fault or fold. From the comparison above, the accuracy of the DEM geological structure analysis's lineament pattern reaches 90% (Fig. 7). This way, it is not burdensome to interpret the structure of the initial stage before doing geological mapping field.

(a) (b)

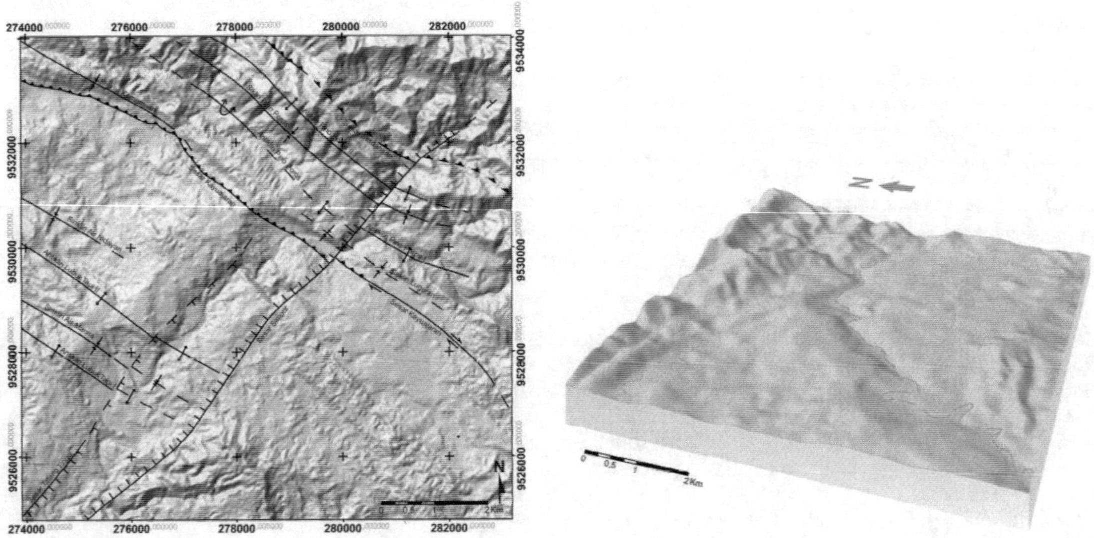

Fig. 7 (a) DEM shows with geological structure ploted, (b) Digital Terrain Model (DTM) shows from 60° bird view direction.

5. Conclusion

The accuracy of geological structures analysis through DEM of DEMNAS reaches 90%. This indicates that the DEM data is sufficiently accurate and beneficial for assisting the interpretation of geological structures prior to mapping on the ground.Those results illustrate the sophistication of technology and the development of science, will facilitate various aspects of human life, including the field of earth science.

The results of this study are expected to be of guidance in interpreting early geological structures and can be applied in relatively similar locations with complex structural zones, with a reflection of topographic control. This study is very important to do in spatial planning and construction of a place, because the geological structure is related to tectonic proceses. The geological structure that is still active, certainly requires special attention in building construction. Other than that, this study can be applied to the industries that have a close relationship with geological structures, such as geothermal or gold mining. Furthermore, this study will be able to be developed again on more specific themes, such as the discussion of validation and accuracy of DEM data.

Acknowledgements

The authors wish to thank Ridho Rizky Amanda, Sahadi, Nopir, Riduan, Sirianto and PT. Bukit Asam (Persero). Tbk., who have supported and sponsored the success of the study and writing and publication of this paper.

References

[1] Amin T.C., Kusnama, Rustandi E. and Gafoer S., 1993 *Peta Geologi Lembar Manna Dan Enggano Skala (1:250.000)*, Geological Research and Development Centre: Bandung.

[2] Badura J. and Przybylsky B., 2005 *Application of Digital Elevation Models To Geological And Geomorphological Studies-Some Exaples*, Przeglqd Geologicnzy, Vol. 53, nr 10/2.

[3] Bolstad P.V. and Stowe T., 1994 *An Evaluation of DEM Acuracy: Elevation, Slope, And Aspect*, American Society For Photogrametry And Remote Sensing, pp. 1327-1332.
[4] Chenrai. P., 2012 *DEM Application for Geological Interpretation: A Case Study at the Koh Samui Area, Gulf of Thailand*, World Applied Sciences Journal, 17(11), pp. 1516-1520.
[5] Fajri S.N., Amanda R.R., Zuhri W. and Sutriyono. E., 2017 *Perkembangan Struktur Geologi Tersier Akhir-Kuarter Pada Bagian Utara Sub-cekungan Manna, Bengkulu*, Prosiding Seminar Nasional AVoER IX, pp. 237-243.
[6] Falorni G., Teles V., Vivoni V., Bras E.R. and Amaratunga K., 2005 *Analysis and Caharacterization of the Vertical Accuracy of Digital Elevation Models From the Shuttle Radar Topographic Mission*, Journal Geophysical Research, 110: F0205.
[7] Guo-an T., Strobl J., Jian-ya G., Mu-dan Z. and Zhen-jiang C., 2001 *Evaluation On The Acuracy of Digital Elevation Models*, Journal of Geographical Sciences, Vol. 11, No. 2.
[8] Gorokhovic Y. and Voustianiouk A., 2006 *Accuracy Assesment of the Processed SRTM-Based Elevation Data by CGIAR Using Field Data From USA and Thailad and its Relation to the Terrain Characteristics*, Remote Sensing of Enviroment, 104, pp. 409-415.
[9] Hall D.M., Buff B.A., Courbe M.C., Seurbert B.W., Siahaan M. And Wirabudi A.D., 1993 *The Southern Fore-Arc Zone of Sumatera: Cainozoinc Basin-Forming Tectonism And Hidrocarbon Potensial*, Proceedings 22nd Annual Convention, IPA, pp.319-344.
[10] Harding T.P., 1973 *Newport-Inglewood Trend, California An Example of Wrench Style Deformation*, American Associoation of Petroleum Geologists Bulletin, V. 57, no.1, pp. 97-116.
[11] Howles A.C.Jr., 1986 *Structural And Stratigraphic Evolution of The Southwest Sumatran Bengkulu Shelf*, Proceedings 16th Annnual Convention, IPA, pp 215-220.
[12] Izart A., Kemal B.M. and Malod J.A., 1994 *Seismic Stratigraphy and Subsidence Evolution of Northwest Sumatra Fore-arc Basin*, Marine Geology 122, pp. 109-124.
[13] Kusnama, Mangga S.A. and Sukarna, D., 1992 *Tertiary Stratigraphy and Tectonic Evolution of Southern Sumatra*, Geological Society of Malaysia-Circum-Pacific For Energy And Mineral Resources Tectonic Framework And Energy Resource of The Western Margin of Pacific Basin, Kuala Lumpur, pp. 143-152.
[14] Ludwing R. and Scheneider P., 2006 *Validation of Digital Elevation Model From SRTM XSAR For Applications in Hidrologic Modeling*, ISPRS Journal Photogramery and Remote Sensing, 60, pp. 339-358.
[15] Matson R.G. and Moore G.F., 1993 *Structural Influences on Neogene Subsidence in the Central Sumatra Fore-Arc Basin*, Geology and Geophysics of Continental Margins, AAPG Memoir 53, pp. 157-181.
[16] Meixner J., Grimmer J.C., Becker A., Schill E. and Kohi T., 2017 *Comparison of Different Digital Elevation Models and Satelite Imagery for Lineament Analysis: Implications for Identification and Spatial Arrangement of Fault Zones in Crystaline Basement Rocks of the Southern Black Forest* (Germany), Elsevier, Journal of Structural Geology, doi: 10. 1016/j. jsg.11.006.
[17] Natawidjaja D.H. and Triyoso W., 2007 *The Sumatran Fault Zone-From Source To Hazard*, Journal of Earthquake and Tsunami, Vol. 1, No.1, pp. 21-47.
[18] Newcomb K.R. and McCann W.R., 1987 *Seismic History and Seismotectonics of The Sunda Arc*, Jurnal of Geopysical Research, Vol. 92, No B1, American Geopysical Union, pp 421-439.
[19] Pulunggono A., Haryo S.A., and Kosuma C.G., 1992 *Pre-Tertiary Fault System As a Framework of The South Sumatra Basin; a Study of Sar-maps*, Proceedings 21st Annual Convention, IPA, pp. 339-360.
[20] Saleem T.A., 2013 *Analysis and Tectonic Implication of DEM-Derived Structural Lineaments, Sinai Peninsula, Egypt*, International Journal of Geosciences, pp. 183-201.
[21] Sieh K. and Natawidjaja D., 2000 *Neotectonics Of The Sumatran Fault, Indonesia*, Journal Of Geophysical Research, Vol. 105, pp. 28, 295-28, 326.
[22] Susilohadi S., Gaedicke C. and Ehrhardt A., 2005 *Neogene Structures and Sedimentation History Along The Sunda Forearc Basins Off Southwest Sumatra And Southwest Java*, Elsevier, Marine Geology 219, pp. 133-154.
[23] Yulihanto B., Situmorang B., Nurdjajadi A. and Sain B., 1995 *Structural Analysis of The Onshore Bengkulu Forearc Basin And its Implication For Future Hydrocarbon Exploration Activity*, Proceedings 24th Annual Convention, IPA, pp. 85-95.

IOP Conf. Series: Materials Science and Engineering **636** (2019) 012002 doi:10.1088/1757-899X/636/1/012002

Transformations of Urban Space in Community and Pedestrian Wellbeing

CMSVillanueva[1]
[1]Assistant Professor 5, Research Fellow, College of Architecture, University of Santo Tomas, Manila, Philippines
E-mail: csvillanueva@ust.edu.ph

Abstract
Resiliency and well-being are pressing issues in communities worldwide as we cope with the challenges of economy, climate change, and the sustainable development goals. Resilience is the ability of ecological systems to absorb changes of state variables, driving variables, and parameters, and still persist (Holling, 1973). The measure of one nation's ability to adapt to these evolving issues may vary greatly with another because of their culture and way of life. The ultimate goal of a resilient society is the wellbeing of its community.

Wellbeing is when individuals have the psychological, social, and physical resources they need to meet a particular psychological, social, and/or physical challenge. When individuals have more challenges than resources, the see-saw dips, along with their wellbeing, and vice-versa (Dodge, 2012).

With two-thirds of the world's population is being expected to reside in cities by 2050, it is clear that urban areas will be the engines of economic growth and sustainable development (UNODC, 2019). Cities across the globe are undermined by chronic insecurity, violence, and corruption, which are often connected to crime challenges originating beyond urban boundaries (UNODC, 2019). Studies show that street crimes are affected by the physical characteristics within a community's field of vision (Lee, et al, 2017).

This study attempts to model the collective experiences of stakeholders and pedestrians from 2015 - 2018. This study will feature the existing conditions on the urban spaces, such as the sidewalks, the road networks, and barangays (barangays are the smallest unit of government in the Philippines). Mapping the social impacts of lighting, noise, crime, and accidents will feature their geographical distribution along España Boulevard; a major access route in the country's University Belt.

Findings show that the manifestations of transformations in urban space occur in fluid trajectories that can be in patterns of horizontal, vertical, and cyclical schemes in achieving community and pedestrian wellbeing.

Keywords:
Community and Pedestrian Wellbeing, Urban Space, Transformations

Methodology
Ambient environmental conditions, such as lighting and noise, were measured with the use of light and sound meters below each lamppost and at every street corner of Manila's España Boulevard, in lux and decibels units, respectively. Agency reports on accidents and crime were collated, tabulated, and mapped to represent the time, day, month, location, vehicles involved, gender, type of crime, and accident. Findings were correlated to reiterate environment and behaviour mitigating circumstances that impact the vulnerabilities in urban space.

Manila's España Boulevard
España Boulevard serves as gateway from the eastern Metro Manila to the country's capital and where most universities are located, forming part of Manila's university belt, more

Content from this work may be used under the terms of the Creative Commons Attribution 3.0 licence. Any further distribution of this work must maintain attribution to the author(s) and the title of the work, journal citation and DOI.
Published under licence by IOP Publishing Ltd

IOP Conf. Series: Materials Science and Engineering **636** (2019) 012002 doi:10.1088/1757-899X/636/1/012002

particularly the University of Santo Tomas. Manila is among the densest cities of the world with a swelling daytime population of more than seventy thousand and more than forty thousand during the evenings (Weller, 2016).

A ground survey was conducted with the help of a light meter and a sound meter installed as smartphone applications to measure both lighting and sound intensities. Surveys were conducted in November 2016 from 6am to 9am, 11am to 1pm, and 6pm and 8pm along España Boulevard. (Villanueva, 2019)

Street lighting was measured during the evenings when lighting intensity was measured below each lamp post, while sound intensity was measured at the corner of each intersection during mornings, noon, and evenings. (Villanueva, 2019)

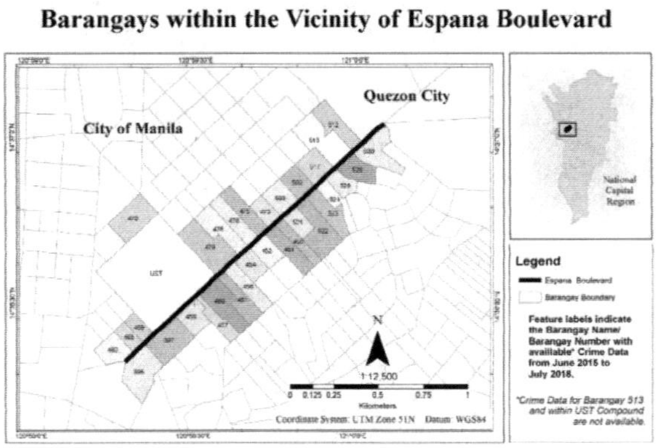

Figure 1. *Map of Barangays within the vicinity of España Boulevard.*
Source: Perez, 2018

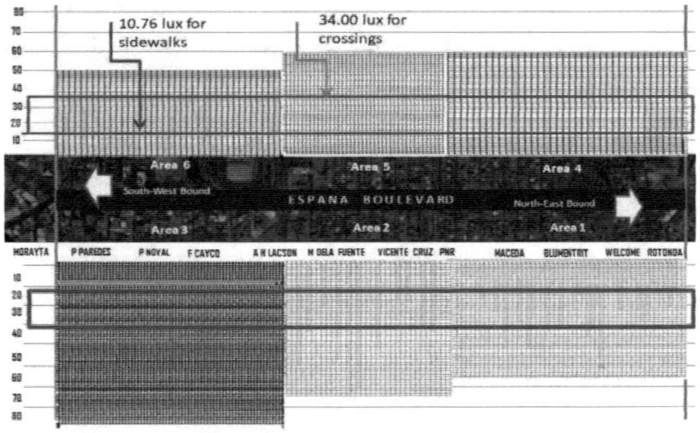

Figure 2. *Illumination of Sidewalks on South-West Bound and North-East Bound of España Boulevard in Lux Units.*
Source: Villanueva, 2019

Figure 2 above shows that España Boulevard was divided into six study areas, with more lampposts and intensive illumination experienced in Area 3, which is across the University of Santo Tomas. This is due to the presence of restaurants, convenience stores, dormitories, and other commercial establishments. (Villanueva, 2019)

Area 5 does not have any lampposts working during the time of the survey and merely shares illumination from commercial establishments across the street and the center island lamps. Billboards also provide much needed street lighting to as much as 80 lux. The required illumination for pedestrians along commercial areas should receive at least 1-foot candle, which is equivalent to 10.76 lux according to the American National Standard Practice for Roadway Lighting. In the Philippines, the Housing and Land Use Regulatory Board (HLURB) standard merely mentions street lighting at fifty meters apart, but does not prescribe the specific required lighting intensity for pedestrian safety and wellbeing. (Villanueva, 2019)

As a main arterial road, connecting Manila to major cities in the national capital region, transport vehicles are a major source of unwanted sound. Urban street noise has been widely assessed as a contributor to a multitude of health issues in different cities around the world. Studies show that exposure to more than 70 decibels for a prolonged period of time has significant implications to human health (McAlexander, Gershon, & Neitzel, 2015; Benfield, 2012).

Manila's Horizontal, Vertical, and Cyclical Challenges in Urban Space

In specifying environmental lighting, the goal is to achieve standards of pedestrian lighting at 10.76 lux average horizontal distance and 34 lux at intersections (McLean, 2006), which has been found to be insufficient based on findings above. Furthermore, incidents of electrocution around the area should accelerate the need to review the specifications of the lamp posts to comply with safety standards. Periodic maintenance should be a cyclical procedure practiced at least every period of two years, which manufacturers of street lighting fixtures are willing to do as a guaranty of quality standards. In other countries, warning signs are also posted vertically on lamp posts, so that pedestrians will avoid direct contact with the street lights, and to mitigate hazards of electrocution.

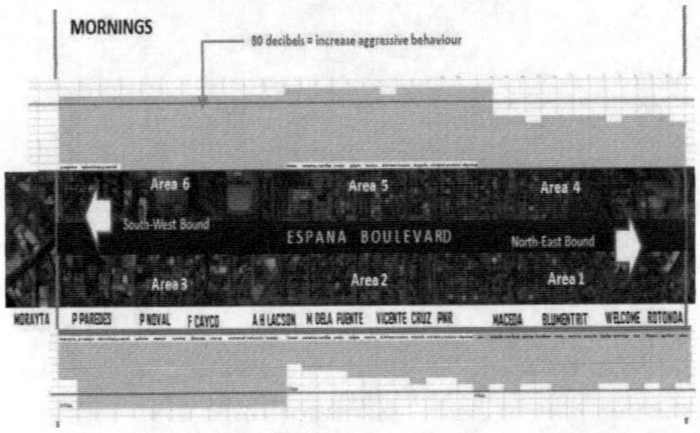

Figure 3. *Sound Levels During a Weekday Survey 6am- 8am in Decibels*
Source: Villanueva, 2019

Figure 4. *Accidents Hot Spots along España Boulevard based on MMARAS 2015*
Image by Rivera, 2017

Studies show that blue illumination from street lights make pedestrians feel safer because they can distinguish facial features of people clearer than those with areas with yellow hues. However, studies also show that there may also be implications with too much glare caused by LED lights, and mishandling may also cause a cycle of emitting toxic fumes.

Tabulated data shows the distribution of noise along España Boulevard, a major thoroughfare that experiences heavy traffic, are critical and a high level of noise along the Rail Road area (Prudencio to Algeciras - Area 5) with as much as 107 decibels in the evening. A.H. Lacson Avenue (a major connector of north to south expressways) during the mid-day and evenings with 100 decibels in Area 6. Trabajo Market area in M. dela Fuente and Ramon Magsaysay High School, experience as much as 109 decibels in evening when the sidewalks are transformed by street vendors to selling areas in Area 2. The bus stop in front of UST can experience as much as 102 decibels in the morning as shown in Area 6.

Goines's and Hagler's studies show that noise levels above 80 decibels "are associated with an increase in aggressive behavior and a decrease in behavior helpful to others". Depending on the duration of exposure, hearing impairment can begin at around 85 dB, "roughly equivalent to the noise of heavy truck traffic on a busy horizontal roadway" (Goines & Hagler, 2007).

The lack of sufficient lighting and too much noise can induce road accidents and mask criminal activities. Correlation of maps can show that insufficient lighting in Area 5 has translated to accidents, crime, and the indifference of the community to help out in these situations.

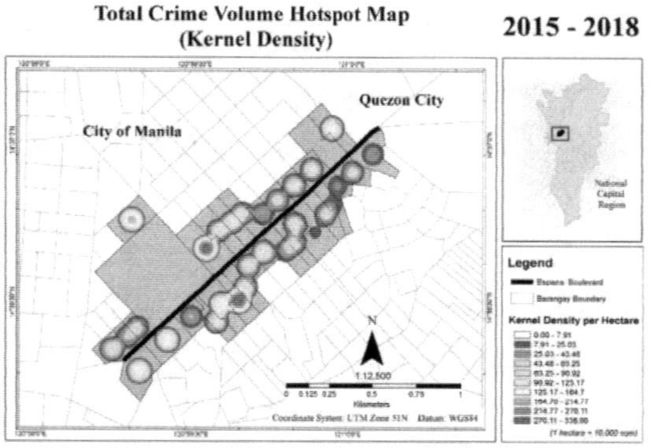

Figure 5. *Kernel density (hotspot) map of total crime volume from 2015 to 2018*
Source: Perez, 2018

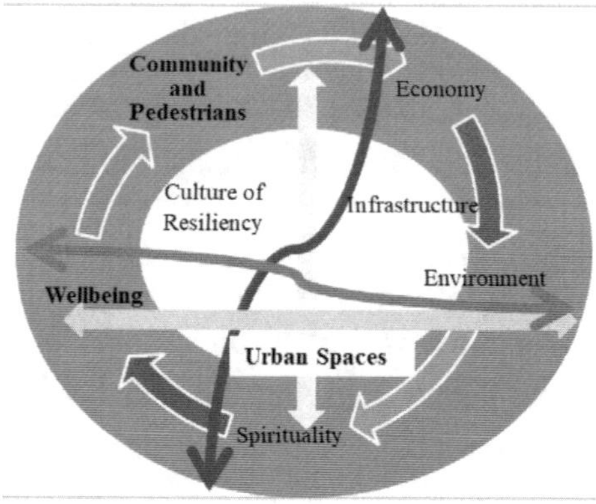

Figure 6. *Transformations of Urban Space in Community Wellbeing*
Source: Author

Manila's Future Resilient Transformations of Urban Space

Asian urbanism has manifested a need to look deeper into the divided views on public space and providing substitutable transformative powers to communities and pedestrians (Demirtas-Milz, 2012). To achieve a cyclical culture of well being, there needs to be connectivity in Manila's perspective, which entails three important guideposts in soft and hard community infrastructure-

1. Build on Manila's cultural strengths;
2. Work for the good of the nation as a whole -acting locally
3. Thinking globally, (De Leon, 2014).

Grassroots effort with initiatives harnessed by the social (horizontal and cyclical) and intellectual (vertical and cyclical) capital of multi-sectoral stakeholders and community planners (Skiba and Dulong, 2008) may manifest in cyclical or non-material factors, such as social supports, freedom, and fairness, which play a bigger role than money in future well-being (Barrington-Leigh, 2018).

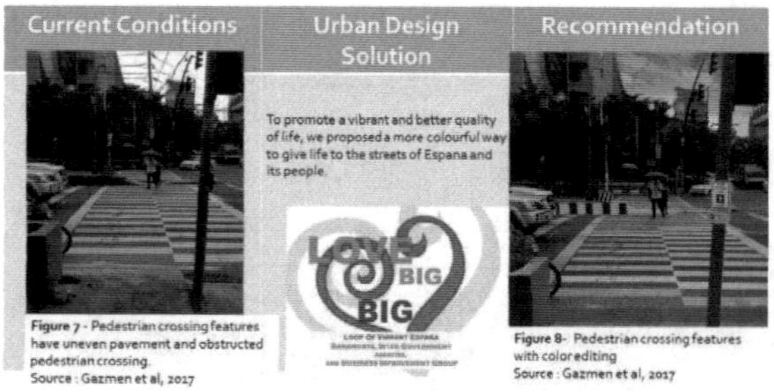

Figure 7-8. *Horizontal Transformations*

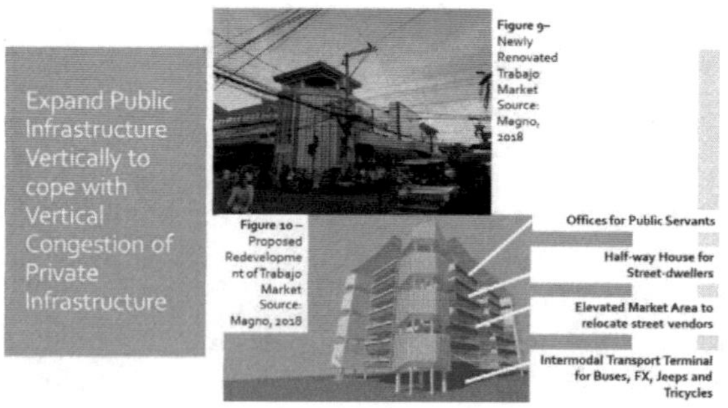

Figure 9-10. *Vertical Transformations*

Figure 11. *Cyclical Transformations*

Figure 12-14. *Cyclical Transformations*

Figure 15. *Cyclical Transformations*

Figure 17-19. *Cyclical Transformations*

We need a cyclical or a more "polycentric", well-connected, and diverse kind of city transformation (Mehaffy, 2019). Cyclical transformations transcend soft and hard infrastructure and focus more on the resiliency of community and pedestrian wellbeing. Humble urban spaces like sidewalks are the arenas of cyclical human interaction, occasionally producing exchanges of information and "knowledge spill overs" that are surprisingly important for vertical economic as well as social development (Jacobs, thru Mehaffy, 2019).

Slowly, the world may change in the direction of leaving no one behind in the common process of the substantive transformation of global societies (Schuerkens, 2017). The vulnerabilities that people encounter in urban space can be mitigated with continued engagement of community movements, the use of mass media, and academic discourses with the youth sector to voice the vision of comprehensive cyclical transformations in pedestrian and community wellbeing.

References:

[1] De Leon, F Jr (2014). A Heritage of Wellbeing: The Connectivity of the Filipino. *Mindanao State University-Iligan Institute of Technology*. Retrieved from https://www.msuiit.edu.ph/ on 20 Feb 2018.

[2] Demirtas-Milz, N (2012). Urban Informality Reconsidered in Neoliberal Context: Gecekondu, Identity, Poverty and Islamic Philanthropism in Turkey. *MacFarlane, C. and Waibell, M. (2012). Urban Informalities: Reflections on the Formal and Informal*. Ashgate Publishing (2012) and Routledge (2016) Abingdon, Oxon and NY, USA.

[3] Dodge, R., Daly, A, Huyton, J, & Sanders, L (2012). The Challenge of Defining Wellbeing. *International Journal of Wellbeing*, 2(3), 222-235. doi:10.5502/ijw.v2i3.4

[4] Goines, L and Hagler, L (2007). Noise Pollution: A Modern Plague. *Southern Medical Journal.* Volume 100, pp. 287-294. Retrieved from nonoise.org/library/smj/smj.htm

[5] Helne, T & Hirvilammi, T (2015). Wellbeing and Sustainability: A Relational Approach. Sustainable Development, *2015 - Wiley Online Library*, Volume 23, Issue 3, May/June 2015. Pages 167-175.

[6] Holling, C S 1973. "Resilience and Stability of Ecological Systems." *Annual Review of Ecology and Systematics 4 (1): 1–23, p. 18) thru Anne Tiernan, Lex Drennan, Johanna Nalau, Esther Onyango, Lochlan Morrissey & Brendan Mackey (2018) A review of themes in disaster resilience literature and international practice since 2012, Policy Design and Practice*, DOI: 10.1080/25741292.2018.1507240

[7] Lee, I, Jung, S, Lee, J, & Macdonald, E (2017). Street crime prediction model based on the physical characteristics of a streetscape: Analysis of streets in low-rise housing areas in South Korea. *Environment and Planning B: Urban Analytics and City Science.* https://doi.org/10.1177/2399808317735105

[8] McAlexander, T P, Gershon, R R, & Neitzel, R L (2015). Street-level noise in an urban setting: assessment and contribution to personal exposure. *Environmental health: a global access science source*, 14, 18. Doi: 10.1186/s12940-015-0006-y

[9] McLean, D (2006) Transport Accident Commission of Australia Roadway Pedestrian Illuminance Levels for Sidewalks. *Guide for the Design of Roadway Lighting*. Retrieved from sunshinecoastastronomy.files.wordpress.com/2015/01/dmd-roadway-lighting.pdf

[10] Mehaffy, M (2019) We Need 'Goldilocks,' Not 'Voodoo,' Urbanism. Retrieved from https://www.cnu.org/publicsquare/2019/01/16/we-need-%E2%80%98goldilocks%E2%80%99-not-%E2%80%98voodoo%E2%80%99-urbanism on February 18, 2019.

[11] Philippine Housing and Land Use Regulatory Board (HLURB), (2009). Revised Rules and Regulations Implementing the Subdivision and Condominium Buyer's Protective Decree, *A Joint Project of the Rules and Standards Development Group and the Technical Working Group on the Revision of the IRR.* Retrieved from http://www.hudcc.gov.ph/sites/default/files/styles/large/public/document/IRR%20PD%20957.pdf on 31 March 2019.

[12] Schuerkens, U, 2017. Social Changes in a Global World. *London: Sage Publications*

[13] Skiba, DJ (1), DuLong D, 2008. Using TIGER vision to move your agenda forward. *Healthcare Informatics, University of Colorado at Denver and Health Sciences Center, School of Nursing, Denver, CO, USA*. DOI: 10.1097/01.NUMA.0000313090.04519.22 . Retrieved http://s3.amazonaws.com/rdcms-himss/files/production/public/FileDownloads/tiger-report-executive-summary.pdf on 20 Feb 2018

[14] United Nations Office on Drugs and Crimes, 2019. Urban Safety and Good Governance. Retrieved from https://www.unodc.org/unodc/en/urban-safety/index.html on 31 March 2019.

[15] Villanueva, C M S, (2019). Community Engagement for Environmental Lighting and Acoustics in the Manila's University Belt. Resourceedings, [S.l.], v. 2, n. 1, p. 140 - 156, Feb. 2019. ISSN 2537-074X. *Available at: https://www.press.ierek.com/index.php/Resourceedings/article/view/457*. Date accessed: 31 mar. 2019. doi:http://dx.doi.org/10.21625/RESOURCEEDINGS.v2i1.457.

[16] Weller, C (2016). Manila is the most crowded city in the world — here's what life is like. *The Business Insider* as retrieved from http://www.businessinsider.com/manila-worlds-most-crowded-city-2016-8 on 16 May 2018.

IOP Conf. Series: Materials Science and Engineering **636** (2019) 012003 doi:10.1088/1757-899X/636/1/012003

Application of parametric components in architectural planning

Wong Leong Yee

Faculty of Architecture, Department of Civil Engineering, Tamkang University, Taiwan

wleongyee@gmail.com

Abstract. The purpose of research is to experiment the use of grasshopper parametric software to analyze a rural district in Taiwan working as planning optimization tools for an aging population community. Conducting an architecture analysis is a process of work mainly dealing with data, as site information getting more complex it is no longer effective carry out by conventional method. Four sets of methodology are performed: (1) Spatial connectivity via road system, (2) path optimization, (3) location optimization for coverage area, and (4) land parcel classification grouping. The experimented results explore a data driven methodology of pre-design analysis in consideration of site context along with residence behavioral aspect, producing a more inclusive investigation for design solution.

Keywords: Parametric tools, site analysis, site planning

1. Introduction

"Technology is the answer, but what is the question?" by referencing the quote from Cedric Price inspire the intention to explore what can architectural planning benefit more from this technology flourishment era. Looking at the pre-design stage, site analysis is carried out to collect information from the soft data (variable measurement such as climate, community sociology and etc.), and also the hard data (permanent physical object containing topography, buildings, road system and etc.). Creating a set of all-covered analysis sometimes results a complex data which difficult to digest for design process. Moreover, the variable aspect collected could not perform instant simulation along with changes when it is presented on a sheet of paper list, which then makes site analysis a more visual aesthetic presentation rather than information driven approach.

Urban dense population issues are attempted to find its solution by numbers of data expert via using algorithm method, articles written available such as "Automated Space Allocation Using Mathematical Techniques" and "Approach to the Multi-Level Space Allocation Problem in Architecture Using a Hybrid Evolutionary Technique" focus on simulating the efficiency residential building for accommodation. Space Allocation process experiment the mix and match of input (habitant) based on specific criteria and ratio in a parameter, which then allows simulation for ideal settings. Using the same ideology of parametric system, the research attempts to apply into a rural community integrating with social and economics aspect in targeting aging community.

Content from this work may be used under the terms of the Creative Commons Attribution 3.0 licence. Any further distribution of this work must maintain attribution to the author(s) and the title of the work, journal citation and DOI.
Published under licence by IOP Publishing Ltd

2. Literature review

In today's advance period of map technology, information beyond geographical aspect such as scene information emerges along with maps, broadening the type of information to cater users need. GIS (Geographical Information System) is one of the tools born under the technological revolution which are widely used in all sectors. Application of GIS in architectural planning provides detailed mapping information that saves a lot of effort in analysis works. In an condition dealing with urban context, the consideration of analysis is more significant as most of the infrastructure is existed, furthermore, the complex subdivision of building plots making zoning and land-used identification a massive amount of works to be carry out. As it is shown how GIS plays a role in architecture field, much 3D modelling software starting to integrate this tool into their system to enhance design process. Plugin such as Meerkat GIS, Heron, and Gismo are written in grasshopper (Rhinoceros 3D parametric tool) as GIS platform for parametric modeling, which makes it as essential tools to be utilized in academia and industries.

3. Methodology

The research conducted specifically in two rural villages at Sanzhi District, Taiwan. It is also one of the districts that having circumstances of aging population, old ages has outnumbered the young due to nation low fertility rate, migration of the youngster into cities for better jobs opportunities simultaneously contribute to the condition. The observed villages have high possibility in turning into a 'ghost town' in 10 to 20 years of time. The researched outcome could provide an analysis base on geographical factors, residences settlement, and social behavior for design planning optimization.

The main source of activity for economy at the villages is agriculture, which had also influencing the social lifestyle of the communities that mostly happen from routine farming, this makes 'population aging' a big impact toward the communities as it reduce social interaction among each other's, further cut down local source of income.

Residential settlement in the area is also much determined by agriculture activities where buildings are sited on irregular and scattered land plot, this created low neighborhood connectivity from one place another. 'Population aging' affects their walkability which also contribute to deterioration social interaction, in addition, road infrastructure in rural is far less efficient compare to the cities which make elderly prefer to stay at home.

3.1. Spatial connectivity via road system

This set of parametric components identifies the connected route from community center (blue circle) to residence preferable gathering spots (red circle) in shortest path. In the situation of low walkability of elderly physique makes activity area shrink closer to their home, hence further away from the community center. Data from the field also shows that the walking distance of elderly (based on various health condition) from home to community center is no longer relevant. The parametric outcome works as a guideline to reconnect both spaces in a more active way by approaching to the elderly, one of the examples could be decentralizing the community center into smaller station based on the frequent used route (shows in darker color).

IOP Conf. Series: Materials Science and Engineering **636** (2019) 012003 doi:10.1088/1757-899X/636/1/012003

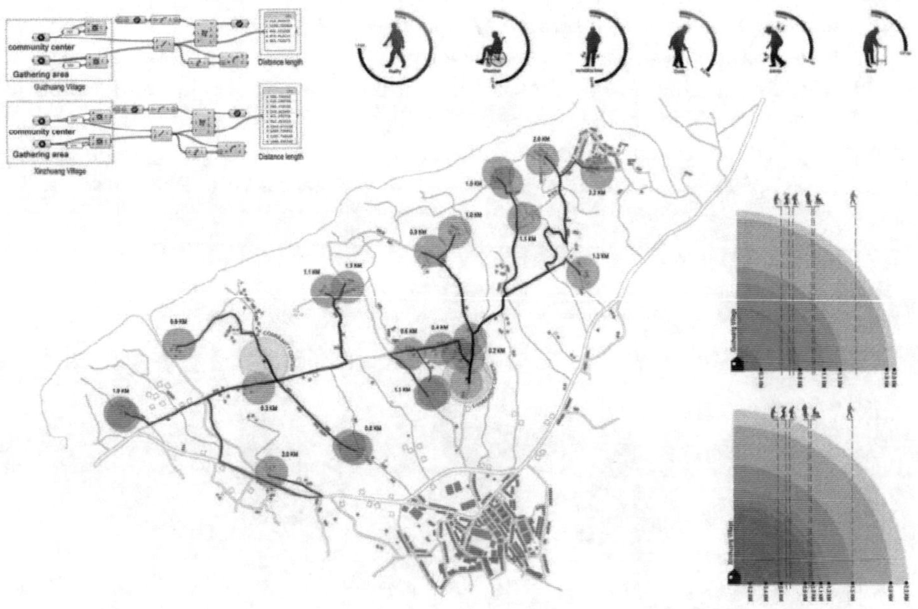

Figure 1: Model of spatial connectivity via road system, various walkability distances based on health condition, and walkability comfort distance based on various locations.

3.2. Path optimization

This set of parametric components picks a random residential unit to identify the shortest route from a specific location spot, along with generating the time estimated for approach according to various form of transport. Medical facilities are a major concern especially in an aged population area, the discontinuous road system is commonly found in rural and agriculture area makes path finding a difficult task especially during emergencies. These outcomes could be further developed to perform as location optimization for service station to cover every residential unit for providing services in shortest time.

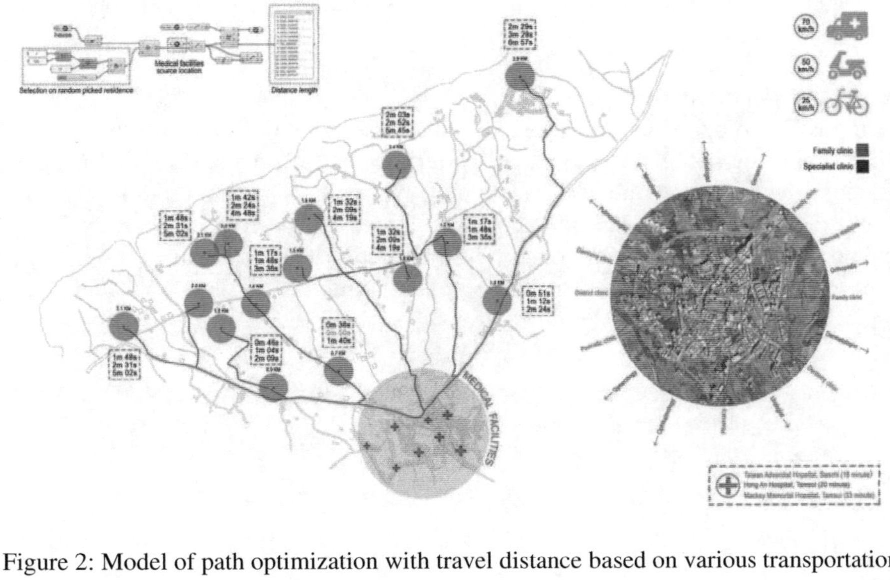

Figure 2: Model of path optimization with travel distance based on various transportation.

3.3. Location optimization for coverage radius

This set of parametric components identifies the coverage radius in targeted spots, temple is the selected as it has large compound area which all has in common, the compound would be activities ground for the community. The model works as finding the uncovered unit to be then reconnect within. Coverage radius was based on sound transmission in referring garbage truck alert music of 70dB, due to the low approachability road system, walking distance will not be considered.

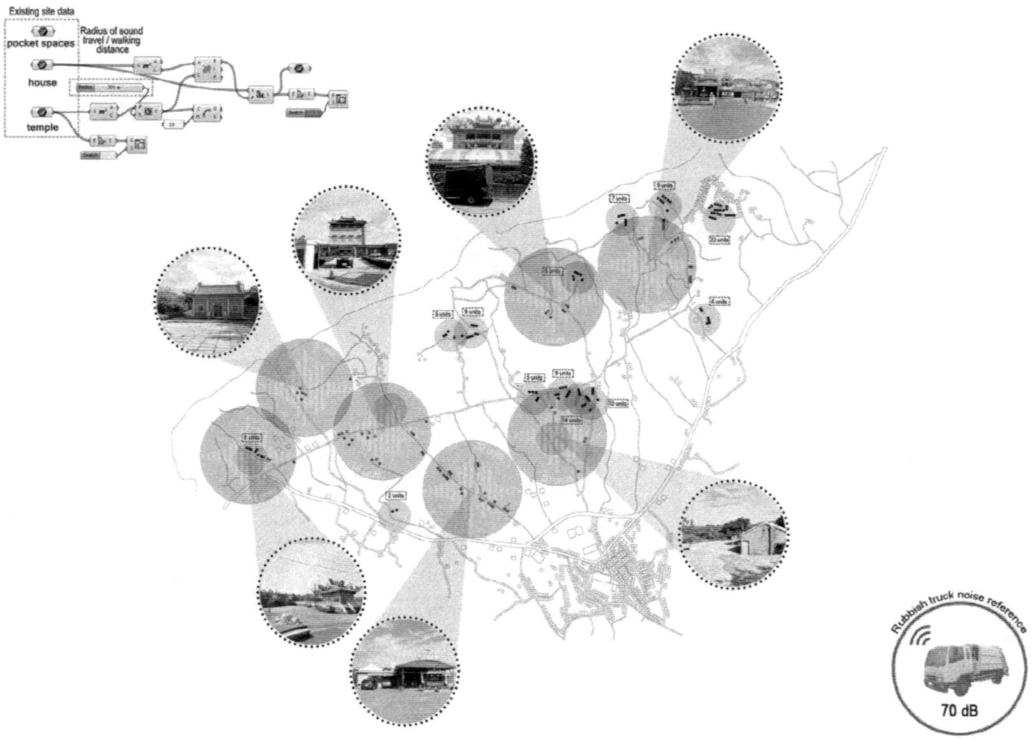

Figure 3: Model of location optimization radius based on sound coverage of 70db.

3.4. Land parcel classification grouping

These both parametric components processes and identify the data of land parcel based on boundary, area size and neighboring parcel. As the parcel pattern in the villages are divided irregularly making difficult to identify the relation with neighboring plots manually. The first model indicate the parcel of agriculture land close to residential boundary (light green) to be used as elderly friendly farming system, simultaneously works as local-consuming production for the villages. Meanwhile the others away from residential boundary parcel (dark green) could be merged and rent for commercial agriculture.

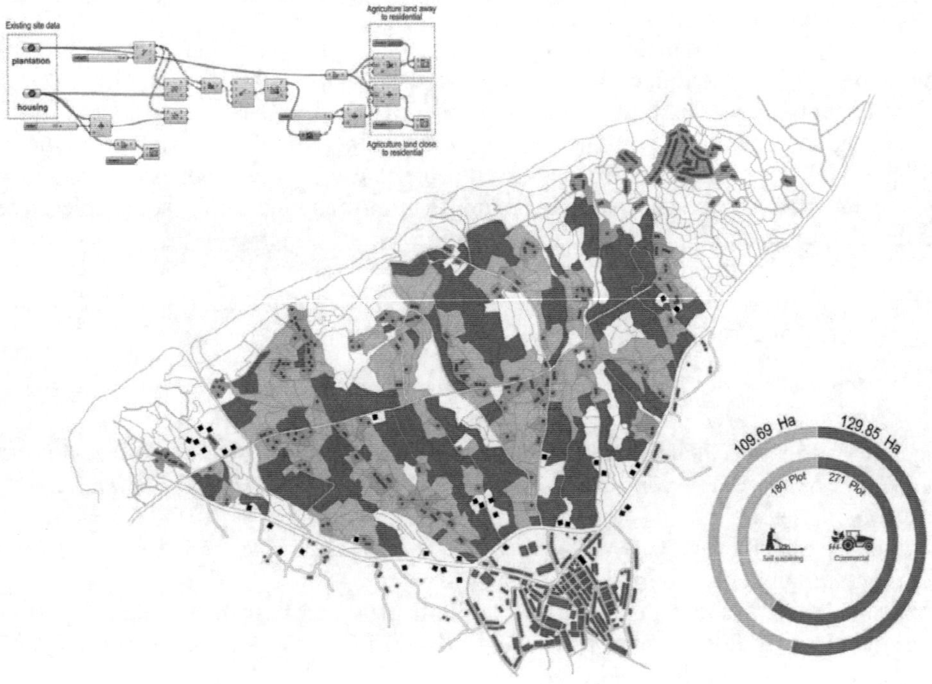

Figure 4: Model of land parcel classification grouping closest to residential boundary

The next model experimented drone driven agriculture system by selecting the smallest land parcel, which are then to be group within an allowed flying range radius to determine the control center location.

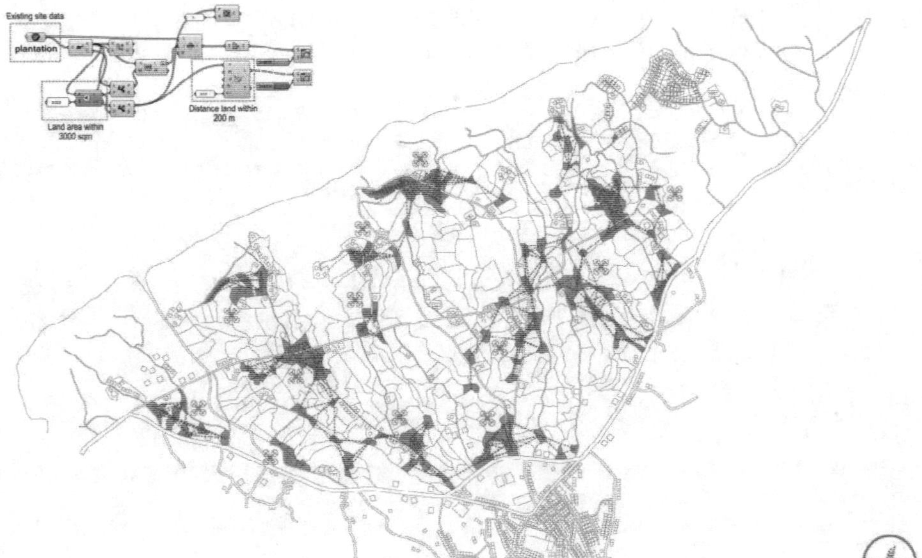

Figure 5: Model of land parcel classification grouping based on land area within 3000sqm.

IOP Conf. Series: Materials Science and Engineering **636** (2019) 012003 doi:10.1088/1757-899X/636/1/012003

4. Conclusion and Discussion

In this research conducted, parametric tools works as a mediator to connect the relations between architectural aspect with economic and sociology that shapes the communities. Grasshopper parametric component allows flexibility in picking random selected location in model (1) and (2) for calculating distance to generate the time needed in approaching, simultaneously merge the generated path pattern in model (2) to maps the frequent used route that works as a potential location proposal. In model (3), parametric components ease the identification of units within adjustable coverage. In model (4), the output instantly performs identification on irregular plot upon smallest plots and nearest plots command. In future researches that expand coverage to others area, the system can be manipulate by adding more inputs like population, road pattern, land plot and etc., constructing a greater network of regional relationship from villages to district and so on.

References

1. Eugenio Rodrigues, Adelio Rodrigues Gaspar, Alvaro Gomes, 2013. *An Approach to the Multi-Level Space Allocation Problem in Architecture Using a Hybrid Evolutionary Technique.* Elsevier
2. Krishnendra Shekhawat, 2015. *Automated Space Allocation Using Mathematical Techniques.* Ain Shams Engineering Journal
3. Mohammadjavad Mahdavinejad, Shaghayegh Shahrigharahkoshan, Mohammadhossein Ghasempourabadi, 2012. *The Role of Site Analysis in Creativity of Student of Bachelor of Architecture, Case: Design Studio III.* Elsevier
4. Valera Sosa, C. Nickl-Weller, 2016. Clinical Research 79 - *Understanding Walkability and Walking rates in Berlin: An Urban Form and Street Pattern Comparison.* Schattauer GmbH
5. Hong-Kun Ni, 2018. *How to deal with agriculture fallow land.* Environmental Information Center
6. Liang-Gong Chen, 2015. *2025 Ageless Era.* CommonHealth publication
7. New Taipei City Sanzhi District Office Accountancy Department, 2015. *2015 Sanzhi District Demographic Analysis.* New Taipei City Sanzhi District Office
8. Frank Stafford, 2014. *GIS in Urban Planning.* GIS Lounge [online] Available at: https://www.gislounge.com/gis-urban-planning/
9. First in Architecture, 2018. *Guide to Architecture Site Analysis.*[online] Available at: https://www.firstinarchitecture.co.uk/architecture-site-analysis-guide-2/

IOP Conf. Series: Materials Science and Engineering **636** (2019) 012004 doi:10.1088/1757-899X/636/1/012004

Invasion of China's Traditional Architecture: A Review

[1]Xia Qing, [2]Mohd Hafizal Bin Mohd Isa*, [3]Muhamad Azhar Bin Ghazali*

[1, 2, 3] Architecture Department, School of Housing, Building and Planning, Universiti Sains Malaysia 11800, Penang, Malaysia

[1]xia1126xia@gmail.com, [2]hafizal@usm.my, [3]azhar.ghazali@usm.my

Abstract. In the process of China's rapid urbanization, the issue of 'uniformity in Architectural identity' has become more serious, while traditional Chinese architectures are replaced by alien buildings. Since 1978, the Chinese government has implemented a series of open policies that have stimulated economic development, driven the development of architecture and changed China's architectural style. Many countries in the context of globalization are inevitably obscured by their own characteristics, especially China. Contemporary Chinese architecture is completely modern and has lost its traditional identity and mainly has Western-style appearances. Chinese traditional architecture has almost withdrawn from the stage of building construction. This paper aims to review the reasons for the current status of Chinese architecture. The formation of China's current construction status has internal reasons in China, and there are also reasons for globalization as the background of China's construction and development. It was found that in recent years, a few architects have begun to pay attention to the construction and development of traditional Chinese architecture. It is hoped that China can continue to develop while not forgetting and abandoning its own traditions, and not losing its own characteristics and identity. The inheritance and development of traditional architecture is also the inheritance and development of culture. Even in the context of globalization, any country should strive not to lose its culture and identity.

1. Introduction

According to [1], the watershed in China's recent history is widely regarded as Deng Xiaoping's 'gaige kaifang' (reform and opening up) project that began in 1978. The tremendous changes in Chinese architecture also began after the implementation of this national policy. China's open policy has brought China into the world economy. The rapid development of China's economy has promoted the development of China's construction industry. This brings China's globalization, especially after China joined the World Trade Organization (WTO) in 2001, [2]. According to [3], trade, together with foreign investment has brought in modern technologies and methods of management that have increased productivity in China, thereby, accelerated the development of China's construction industry. This also has increased the migration of expatriates, which have joined the construction industry in China. Under the guidance of Western architectural trends, China's construction industry has greatly improved the design level, and has also achieved some remarkable development achievements. However, there have been some drawbacks in the process of development. Based on a study done by [4], the home-building promotion program has resulted in a large number of residential buildings with small windows and high-rise buildings with large glass.

Content from this work may be used under the terms of the Creative Commons Attribution 3.0 licence. Any further distribution of this work must maintain attribution to the author(s) and the title of the work, journal citation and DOI.
Published under licence by IOP Publishing Ltd

IOP Conf. Series: Materials Science and Engineering **636** (2019) 012004 doi:10.1088/1757-899X/636/1/012004

Figure 1. Alien architecture is replacing traditional architecture

In addition, more and more alien architectures are replacing traditional Chinese buildings, [5]. Alien architecture is eating buildings with Chinese characteristics like a huge hungry crocodile (Figure 1). In terms of building materials, reinforced concrete is more preferable compared to traditional building materials, such as natural stone, bamboo, wood, and clay. In terms of building mass, large-scale skyscrapers replaced low-rise buildings. Exquisite garden architecture, ethnic architecture, or religious architecture has not been seen in the construction of buildings in recent years. Most of the facades of modern Chinese architecture are glass curtain walls. Whether you live in the North or South of China, coastal or inland areas, the architecture of the cities are very similar, which do not portray the importance of sense of place. Therefore, this study proposes that the future of architecture should be planned wisely and rationally. Alien architecture should not completely replace traditional Chinese architecture. Chinese cities should have their own characteristics, and architecture should emphasize regionality. This paper analyzes the reasons for the formation of contemporary Chinese architecture.

2. Internal reasons for the current situation of Chinese architecture

The development and change of Chinese architecture is influenced by China's modern history. In 1840, the beginning of the Opium War marked the end of the feudal dynasty that started to affect the traditional Chinese architecture. During this period, the Chinese people were at war and struggled to fight, and knowledgeable Chinese began to accept and learn Western technologies to improve their 'combat abilities'. Wei Yuan proposed to learn from the advanced technologies in the West in order to resist the invasion of the Western powers. The idea of this concept laid the foundation for learning from the West and played a positive role in China's development. In 1949, the People's Republic of China was established. The country after the war was facing a major task of post-disaster reconstruction. However, the development of the country had not started and carried out in a positive way because of the wrong plans and implementation in national policy. However in 1978, Deng Xiaoping began to implement the reform and opening up policy during his leadership at that time. According to [1], the policy has boasted up economic development and material prosperity. Before the implementation of the reform and opening up policy, no matter whether it was Beijing or Shanghai, almost no tall buildings could be seen. After the implementation of a series of open policies in China, 40 years have passed, and China's urban constructions have taken on a new shape and look. Not only large cities such as Beijing and Shanghai, but even small cities in tier 3 or tier 4 are filled with tall buildings, [4]. In addition, managers and developers are increasingly passion about new technologies, [6]. Under the guidance of advanced Western technologies and ideas, Chinese architecture has developed vigorously. The emergence of new materials has led to the emergence of alien building types and architectural forms that are out of the harmony of existing urban environments in China, [6]. There are three reasons for the replacement of traditional Chinese buildings by alien buildings.

2.1. Cultural inferiority

According to [7], the mainstream of Chinese officials believes that past traditions have led to China's aggression, and that Western learning has allowed China to develop. It is inevitable and necessary to say goodbye to past traditions. Therefore, starting from the reform and opening up policy in 1978, the development of Chinese cities reflects a top-down identity split. This is the main reason for the abandonment of traditional culture. It seems to be blamed on the national inferiority caused by the war in China. That is what the ancestors were not good at, whether it was political system or culture, which was the reason why China was invaded. On the contrary, as long as it is foreign (Western), it is good. After the implementation of the reform and opening up policy, the huge visual gap between China and the West has greatly enhanced this feeling of inferiority. This prompted us to say 'goodbye' to the past quickly. Moreover, Chinese leaders and officials mainly focus on the economy and ignore the inheritance and development of the national culture, [7]. China's urbanization has changed patterns of daily life and urban environment, which has led to the loss of architectural heritage, [6]. New political and social formations, new sets of values and beliefs, new social identities, and new subjectivities have emerged, [6]. People in urban areas are often satisfied with Western-style of architecture. But gradually, the sense of freshness and modernism will be replaced by the sense of loss of identity. Architecture is the embodiment of culture. Losing the traditional architectural culture contributed to the loss of its identity to a certain extent.

2.2. Time constraint

It was found by [8] and [9] that the level of urban population in China was less than 20 percent of the population in 1978, and then reached 59.58% in 2018. The rapid growth of urbanization in the past 40 years is due to a series of national policies. Because of the wrong policies in the early days of the founding of the country, the construction of the country has been in a blank period. Therefore, in the initial stage of the implementation of reform and opening up, the government has only one requirement for construction, which is fast. This made the architects, who were not ready to build a new Chinese city at that time, have no time to think. They began to choose the simplest and fastest way to copy the urban architecture of western developed countries, so that China also has many modern cities. Based on a study done by [4], Chinese local governments encourage building to boost local output and employment, as well as to raise revenue from land sales. The government's policy is to let the Chinese construction industry blindly pursue and imitate the modern development of Western developed countries. The culture of traditional Chinese architecture has not been effectively passed down and developed. Under the government's promotion policy, people are busy building high-rise buildings, but they have forgotten their traditional buildings. In the initial stage, China chose Western-style architecture to abandon traditional architecture. After 40 years, the construction of Chinese architecture seems to be still imitating the copying stage, and there is no awareness of architectural innovation and improvement. In the process of construction, the problem of 'one thousand cities have one face' has gradually emerged. Clearly, a one-size-fits-all approach to strategic planning for urban development in China is untenable, [8]. However, the decision-makers of China's construction industry still have no effective solutions to this problem, and there is little interest in the inheritance and development of traditional architecture.

2.3. No traditional architectural aesthetic awareness

According to [5], contemporary architectural practices lack the sensitiveness towards the natural environment. For China, the practicality and modernization of architecture is the primary choice. That is to say, as long as the building can be used, the architect does not have to think about the aesthetics of the building. Therefore, until today, the understanding of architecture is still available, from the builders of the building to the users of the building without an aesthetic sense of the exterior of the building. Of course, the most important reason is that those who decide the style of architectural design have a preference and blind worship for Western architecture. The one-size-fits-all model mentioned above is also a manifestation of the lack of architectural aesthetics, [8]. Building developers

IOP Conf. Series: Materials Science and Engineering **636** (2019) 012004 doi:10.1088/1757-899X/636/1/012004

demand that architects design buildings that are more international in appearance and closer to Western modern architecture. Sometimes they even ask for some exaggerated or even strange buildings. However, the space and form of buildings are the most frequently constituted of society. Their visual image announces its presence – in the city, in the nation, and in various distinct worlds, [10]. Just like the dressing style of people, national costumes are neglected or even abandoned. The mainstream of the market is the so-called 'internationalized' products. The current objective facts are that large cities, such as Shanghai and Beijing, have developed at a comparable level to the West, and urban imagery tends to be homogenized with Western cities. Chinese cities have lost their brands in the aesthetics of architecture.

3. External reasons for the current situation of Chinese architecture

Few people pay attention to the inheritance and development of traditional Chinese architecture. It was found by [11] that the phenomenon of globalization and the formation of new global networks, both physical and virtual, impact upon our perception of both place and identity. In today's world, the rise of a global information society has also led to foreseeable changes. The 'cultural convergence' brought about by 'globalization' will inevitably lead to 'construction convergence' between regions. Based on history, special culture and experience that used to be the background of traditional architectural significance in the past have been ruled out by globalization. The connection between architecture and a particular society is gradually blurred. A common criticism of globalization is that it tends toward uniformity, and that it undermines local cultures and value systems, [8]. The globalization of technology and production methods has brought about the separation of human and traditional geographical space, and the characteristics of regional culture have gradually declined. Standardized production of goods leads to a convergence of the building environment, mediocrity of design, and reduction of the diversity of architectural culture.

After series of China's open policies, they have contributed towards China's globalization. Especially after joining the WTO in 2001, advanced construction technology and experience, advanced architectural ideas, and sufficient funds have accelerated the rapid development of China's construction industry. While driving the rapid development of China's economy, it also injected a different culture into China. According to [7], the assessment that often resulted was that globalization posed a potential threat to China. The biggest impact of globalization on China's cities and buildings is the rapid development of urbanization and the major projects in urban planning and architectural design. Most of the projects are led or participated by international architects. The exchange and integration of world cultures have made culture more unified. A critical question is whether the dispositions associated with Chinese culture can survive the interjection of foreign modes of expression, [8]. For China, globalization is an opportunity, but a challenge, 'double-edged sword' views of globalization remains dominant in China, [7]. The mainstream view of Chinese leaders, officials, and researchers is that globalization poses dangers but that it is inevitable and that China has no rational choice other than to fully participate in the process, [7]. The Westernization of Chinese architecture seems to be an inevitable trend. But future developments are still influenced and determined by government policies. In fact, most countries are facing the challenges of globalization like China. Therefore, a correct national policy is really essential.

4. Discussion and Conclusion

This paper reviews the reasons that affect the status quo of Chinese architecture. As mentioned above, China's economy has developed rapidly for nearly half a century, more and more alien buildings have sprung up. Some designers and construction investors have done some exaggerated and strange architectural appearance to attract people's attention. Few people pay attention to the inheritance and development of traditional Chinese architecture.

Architecture is a large-scale embodiment of human culture. Architecture is not a work of art that needs to be seen in museums, like calligraphy, painting, or sculpture. Architecture is a kind of 'artwork' that you can enjoy as long as you are willing to see it. As a manifestation of culture, architecture should retain its own geographical features. Architecture is a symbol of social culture, and its visual image often declares its place of existence. Chinese architect Wang He said that architectural culture should emphasize regionalism. China's traditional culture has been completely destroyed along with the rapid development of the economy and the decline of traditional morality. The traditional architecture, as the carrier of Chinese culture is also devastated by the cultural catastrophe and the rapid urbanization tide.

The building reflects the development of a civilization. Traditional Chinese architecture is a ship that carries information about specific social values. The values we hold now are affecting the course of the ship. As a kind of existing civilization, whether civilization can be inundated by Western civilization in modern society, depends not only on the continuation of the age or the influence of radiation, but also on whether this civilization can truly stand in harmony and integration. If Chinese architecture continues to blur its regional cultural characteristics or even abandon it completely, and continue to build alien architecture without paying attention to traditional Chinese architecture, then Chinese architecture may lose itself, no longer have its own characteristics, and lose its Chinese brand. China should think critically about globalization and improve its ability to create and innovate to meet the challenges of globalization without losing identity.

Acknowledgements

The authors would like to thank the School of Housing, Building and Planning, Universiti Sains Malaysia.

The authors gratefully acknowledge the financial support of Research University Grant (RUI) (Ref No: PPBGN1001/PPBGN814285).

References

[1] Liu K 2004 *Globalization and cultural trends in China*
[2] Dong X, Bowles P and Chang H 2010 Managing Liberalization and Globalization in Rural China: Trends in Rural Labour Allocation, *Income and Inequality Globalization and Labour in China and India* (London: Palgrave Macmillan UK) pp 25–44
[3] Chow G C 2005 *Globalization and China's Economic and Financial Development*
[4] Glaeser E, Huang W, Ma Y and Shleifer A 2017 A Real Estate Boom with Chinese Characteristics *J. Econ. Perspect.* **31** 93–116
[5] Senarath N K 2001 *Design with the Nature*
[6] Cetin M 2010 Trends of the Transformation of Urban Form in Arab Cities with Specific Reference to Al-Khobar *J. Civ. Eng. Archit.* **4** 51–7
[7] Garrett B 2001 China Faces, Debates, the Contradiction of Globalization *Asian Surv.* **41** 409–27
[8] Heikkila E J 2007 Three questions regarding urbanization in China *J. Plan. Educ. Res.* **27** 65–81
[9] Xiru L 2019 The total population has grown steadily, and urbanization has been steadily increasing. Bur. Natl. Stat. China
[10] Anthony D.King 2004 *Spaces of Global Cultures* vol **84**
[11] Brown R and Maudlin D 2011 Concepts of Vernacular Architecture 340–55

IOP Conf. Series: Materials Science and Engineering **636** (2019) 012005 doi:10.1088/1757-899X/636/1/012005

Long-Range Transport and Local Emission of Atmospheric $PM_{2.5}$ in Southern Region of Peninsular Malaysia

N Dahari[1], K Muda[1*], N Hussein[1], MT Latif[2], MF Khan[3] and MS Mohamad Khir[1]

[1]Faculty of Civil Engineering, Universiti Teknologi Malaysia, 81310 Johor Bahru, Malaysia

[2] School of Environmental and Natural Resource Sciences, Faculty of Science and Technology, Universiti Kebangsaan Malaysia, 43600, Selangor, Malaysia.

[3] Department of Chemistry, Faculty of Science, Universiti Malaya, 59100 Kuala Lumpur, Malaysia

Corresponding author*: khalida@utm.my

Abstract. The excessive inhalation of pollutant particles with aerodynamic diameters less than 2.5 μm ($PM_{2.5}$) results in risking the human respiratory health. The aim of this study is to investigate the possible pathways of long range and local tranportation for $PM_{2.5}$ sources. The route patterns are analyzed by using the Hybrid Single-Particle Lagrangian Integrated Trajectory (HYSPLIT 4.9) software during August 2017 to January 2018, covering three main monsoon seasons of Malaysia; southwest (SW), intermonsoon (IM) and northeast (NE) monsoon. This study represents the 72 h air mass backward trajectories in a mixed commercial-residential-industrial site of Skudai, Johor Bahru. The identified source origins are largely varied with different monsoon seasons and meteorological factors. The air-flow at different altitudes results in different effects, as the trajectory at height 500 m contributes to the greatest impact on the pollution level. The potential sources were generally from the Sumatera, Indonesia, during SW monsoon and China, during NE monsoon. The $PM_{2.5}$ exceedance pattern in the Skudai area is also strongly influenced by the localized circulation, anthropogenic emission and dispersion characteristics. The increased levels of $PM_{2.5}$ are predominantly contributed by the primary sources of vehicular emissions and industrial activities.

1. Introduction

One of the main pollutants that contributes to the negative impact of the global climate is airborne particulate matter (PM) [1]. $PM_{2.5}$ pollutant can reside for a longer period of time in the atmosphere. The excessive exposure to $PM_{2.5}$ can pose severe health problems due to its capability of penetrating deep into the alveolar region of human respiratory system and causing various diseases including heart attack, acute bronchitis, asthma and lung cancer [2]. Location, background area, gaseous pollutants and seasonal meteorological factors influence the composition and chemical variability of the fine

Content from this work may be used under the terms of the Creative Commons Attribution 3.0 licence. Any further distribution of this work must maintain attribution to the author(s) and the title of the work, journal citation and DOI.
Published under licence by IOP Publishing Ltd

particles [3]. Source apportionment is a method that can quantify the sources of chemically speciated particles including $PM_{2.5}$ pollutant in order to ascertain the sources responsible for each particle emission [4]. The approach of determining the emission sources involves working backwards after collecting $PM_{2.5}$ samples at the required receptor [5].

2. Methodology

2.1. *Study Area*

The location is selected at southern part of Malaysian Peninsula with coordinates, 1°33'54"N 103°38'24"E (1.565, 103.64), located at Universiti Teknologi Malaysia (UTM) Skudai, Johor Bahru, in order to study the pathways of the ambient air. The study was conducted during 6 consecutive months to cover southwest (SW; August 2017-September 2017), inter-monsoon (IM; October 2017-November 2017) and northeast monsoon (NE; December 2017-January 2018) of Malaysia.

2.2. *Air-Mass Backward Trajectory Analysis*

Hybrid Single-Particle Lagrangian Integrated Trajectory (HYSPLIT 4.9) software model of National Oceanic and Atmospheric Administration's (NOAA) Air Resources Laboratory (ARL) was used to generate three dimensional (3-D) plots, also, to calculate findings and air mass trajectories based on metrological parameters [6]. This approach is to investigate the source origin and transport patterns of the pollutant emission at the sampling sites [7]. The approach was done at the initial step of evaluating the ambient particulate concentration prior to determining the emission source. To analyse seasonal variations, HYSPLIT was computed each month from August 2017 to January 2018 using the meteorological data obtained from the Global Data Assimilation System (GDAS) provided by the National Centers for Environmental Prediction (NCEP). Since the $PM_{2.5}$ has the ability to travel long distance, thus, the 72 h air-mass back-trajectories for wind direction were calculated every 6 h a day (00, 06, 12 and 18 h UTC) onto the plot simulation diagram. Trajectory data of multiple altitudes of 500 m, 1000 m and 1500 m above ground level (AGL) represented the air masses arriving at UTM Skudai.

3. Results and Discussions

Figure 1 shows the pathways of the air mass patterns during the seasonal periods of southwest (SW), intermonsoon (IM) and northeast (NE) monsoon, while Figure 2 displays the zoomed-in route patterns of backward air trajectories covering the three monsoon seasons. The airstreams during SW are mainly maritime as they are usually drawn from the western Pacific and from the Indian Ocean, while the air during NE is normally supplied by the regions of Northeast Asian and Australian continents. Based on Figure 1(a) and Figure 1(b), the long range air mass is found to be originated from the regional source of Sumatra, Indonesia, particularly during the southwesterly monsoon. Based on Figure 1(a) and Figure 1(b), the long range air masses are found to be originated from the regional source of Sumatra, Indonesia, particularly during the southwesterly monsoon. The boundary layers of altitudes 500 m, 1000 m and 1500 m seem to also originate from the same region. During August, the airmass trajectories of different altitudes started to traverse northwest of Java, directed towards Singapore, before shifting to Skudai. Although, when the long range tranported pollutants are taken into account, the source was far from Malaysia, but the prevailing airstream only took 3 days to arrive to Skudai. Hence, the particular $PM_{2.5}$ mass concentration was the highest during the sampling duration period of three monsoon seasons. The source of airmass during this monsoon is from southern Sumatra and stretches to the end of Jakarta. Its diurnal upslope position flows towards the neighbouring landmasses.

Locally, during the SW monsoon of August 2017−September 2017 (Figure 2(a) and Figure 2(b)), the $PM_{2.5}$ pollutant is observed to be emitted from the industrial area of Gelang Patah and Johor Bahru

IOP Conf. Series: Materials Science and Engineering **636** (2019) 012005 doi:10.1088/1757-899X/636/1/012005

City Centre, and a few trajectories directed from the Strait of Johor. All the trajectories were pointed from the same direction. Since the SW monsoon is still well developed during this period, the airstream is southeasterly, it results in the rising of $PM_{2.5}$ mass concentration in Skudai. By the end of SW monsoon time, the SW winds started to weaken and move the southern boundary further south to Peninsular Malaysia, affecting the $PM_{2.5}$ mass concentration in the study area. The $PM_{2.5}$ peaks were due to the other meteorological conditions, and collided with the shift in the wind patterns. It can be seen that the source of emission is directed north towards Skudai, through Singapore and Johor Bahru centre. Moreover, the proximity of Johor Bahru and Singapore to these sources enhances the region to be more susceptible to the pollutants, hence, the rising of its concentration during this period. In addition, the increase of pollution level in the SW monsoon can also be attributed by nearby anthropogenic and local sources, as stated above. At this time, the high $PM_{2.5}$ concentration in Skudai was favoured by the surges of prevailing wind system and the local pollution, suggesting that the worse air pollution was due to transboundary airmass transportation and nearby emissions [8].

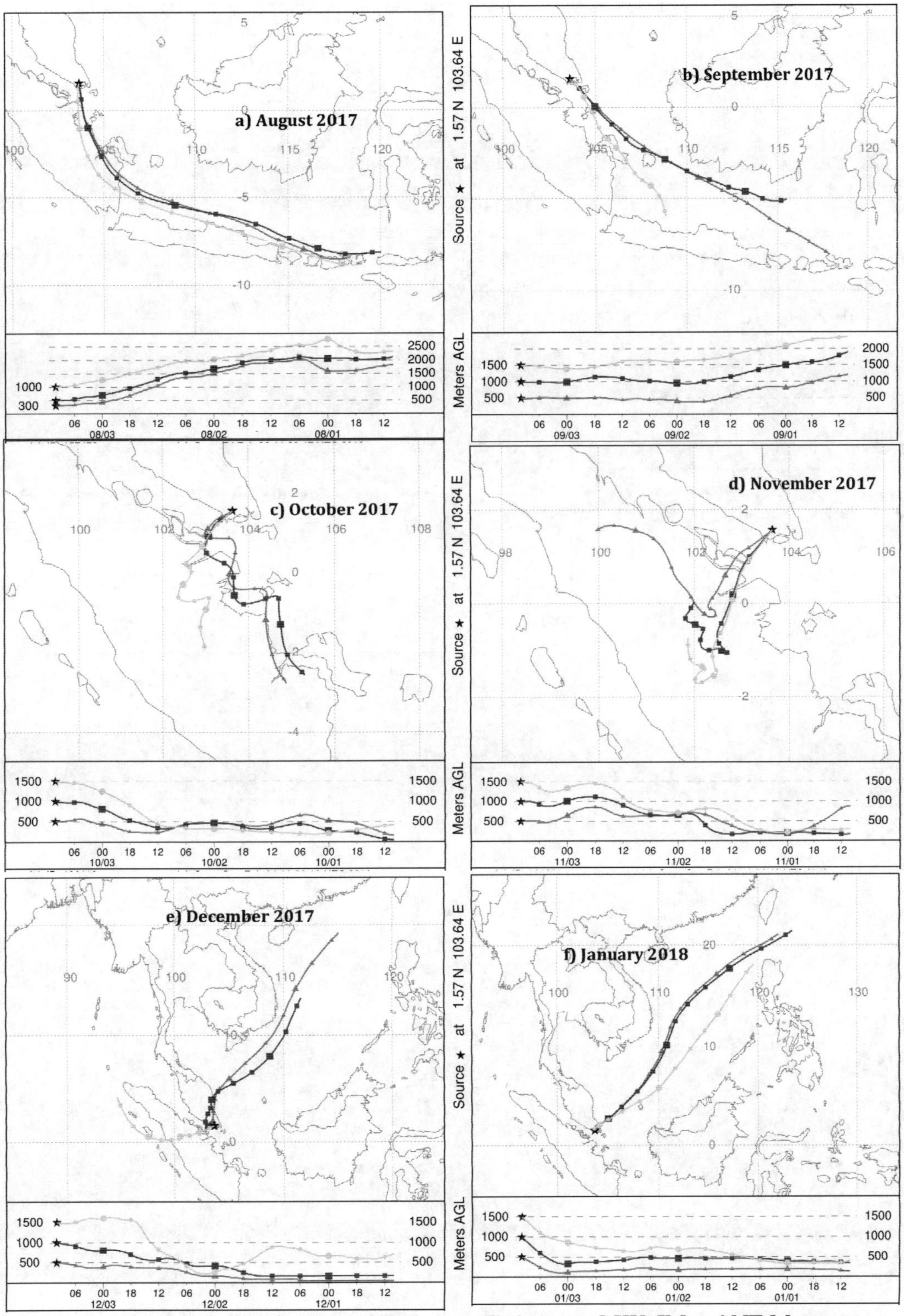

Figure 1. 72 h Backward Trajectory Pathways of SW, IM and NE Monsoons

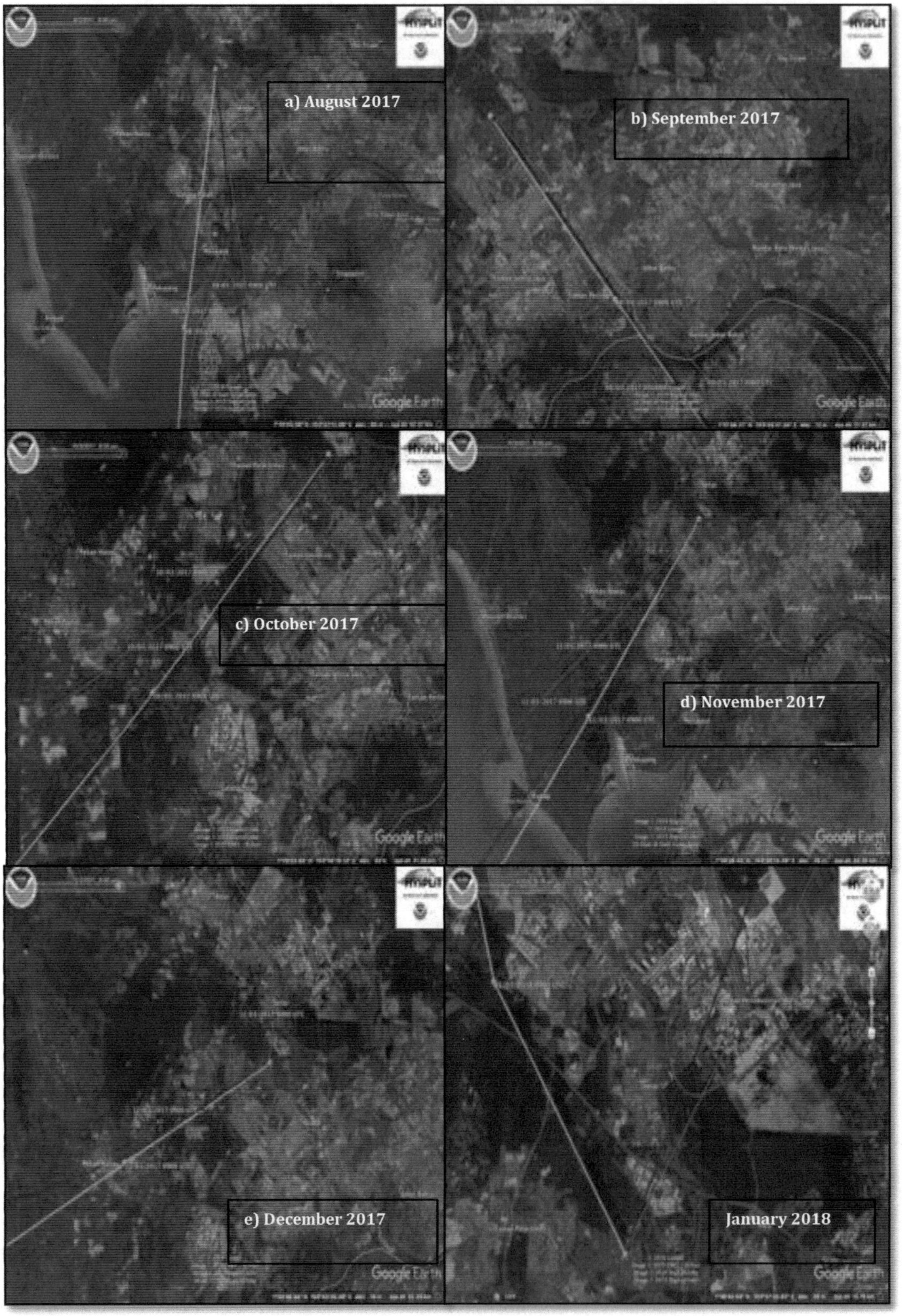

Figure 2. 72 h Backward Trajectory Pathways of SW, IM and NE Monsoons (Zoomed-In)

By the early October, the SW monsoon was starting to fade as the winds were varying and lowering during this duration. Later, the inter-monsoon started to signal as the Northern Equatorial Boundary was observed to move towards the south. The wind field during the intermonsoon season (IM) shows that the wind blows from west Sumatra (Figure 1(c)). However, during November 2017 (Figure 1(d)), the lowest altitude of airmass had a different pathway pattern compared to the other two altitudes, as the lowest mixing layer is observed to be transported in a long range originated from North Sumatra, while others are from Central Sumatra. The winds and the height of the mixing layer that were representated by the lowest layer (active boundary layer) had great influence on the concentration of $PM_{2.5}$ [9]. Since the single height analysis had large uncertainties [10], the analysis consisted of the well-mixed convective boundary layer of low, middle and high atmosphere layers which had different characteristics on the $PM_{2.5}$ concentrations [11].

During the IM monsoon period, the winds are normally lighter. Figure 2(c) and Figure 2(d) display that the pollutant is transported from Pekan Nenas and Pontian during the IM period. In the last days of the transition monsoon, the $PM_{2.5}$ mass concentration in the study area was already decreasing, suggesting significant findings contributed by the PM mass and seasonal variations. The weak winds enhanced the aerosol transboundary transportation from neighbouring region towards the southern region of Peninsular Malaysia. By the end of IM monsoon, the trajectories started to shift northeast as in Figure 2(e) and Figure 2(f), which show that the NE monsoon winds start to develop in the far north of Pacific, slowly advancing to move southwards through the eastern China Sea arriving at the study area (SEA). The trajectories in the starting of NE monsoon of December were dispersed, directing southwards. As the boundaries are well developed, the monsoons dominate the circulation over a region. The concentration of $PM_{2.5}$ is largely dependent on the direction of the prevailing winds and the intensity of anthropogenic emissions localized in Senai and Pekan Nenas.

4. Conclusions

In this study, the $PM_{2.5}$ backward trajectories suggested the high possibilities of pollutants being transported in a long range from the highly polluted regions that could significantly influence the air quality in the less polluted areas. However, it is undoubtable that the local emissions also affect the $PM_{2.5}$ concentration due to the results of the study area which was located in a mixed residential-industrial-commercial environment. It is clear that despite the anthropogenic activities as the sources of particulates, a minor fraction of pollutants is also due to the regional transboundary transport. This valuable information is essential to prevent the likely health threats to the community of Skudai-Iskandar developed region since the $PM_{2.5}$ pollutant was locally generated, predominantly from vehicular emissions. In addition, the findings also suggest that the emission of primary pollutants was favoured by the proximity of the site to the Skudai Highways, Johor Bahru city centre and Senai industrial areas, suggesting probable source origins of mobile and industrial emissions that contribute to the variations of $PM_{2.5}$ mass concentrations. In conclusion, the local authorities should encourage the use of public transportation to improve the state of air quality at local and regional levels.

Acknowledgements

This research is assisted by the financial support from the research university FRGS grant (R.J130000.7822.4F984) from Universiti Teknologi Malaysia, Skudai and GUP TIER II grant (Q.J130000.2622.14J61, Q.J130000.2722.02K82 and Q.J130000.2622.02J54)

References

[1] Mallet M, Dulac F, Formenti P, Nabat P, Sciare J, Roberts G, Pelon J, Ancellet G, Tanré D, Parol F, Denjean C, Brogniez G, di Sarra A, Alados-Arboledas L, Arndt J, Auriol F, Blarel

L, Bourrianne T, Chazette P, Chevaillier S, Claeys M, D'Anna B, Derimian Y, Desboeufs K, Di Iorio T, Doussin J-F, Durand P, Féron A, Freney E, Gaimoz C, Goloub P, Gómez-Amo JL, Granados-Muñoz MJ, Grand N, Hamonou E, Jankowiak I, Jeannot M, Léon J-F, Maillé M, Mailler S, Meloni D, Menut L, Momboisse G, Nico- las J, Podvin T, Pont V, Rea G, Renard J-B, Roblou L, Schepanski K, Schwarzenboeck A, Sellegri K, Sicard M, Solmon F, Somot S, Torres B, Totems J, Triquet S, Verdier N, Verwaerde C, Waquet F, Wenger J and Zapf P 2016 Overview of the Chemistry-Aerosol Mediterranean Experiment/Aerosol Direct Radiative Forcing on the Mediterranean Climate (ChArMEx/ADRIMED) Summer 2013 Campaign *Atmos. Chem. Phys.* **16** 455–504

[2] Gao JJ, Tian HZ, Cheng K, Lu L, Zheng M, Wang SX, Hao JM, Wang K, Hua SB, Zhu CY and Wang Y 2015 The variation of chemical characteristics of PM2.5 and PM10 and formation causes during two haze pollution events in urban Beijing, China *Atmos. Environ.* **107** 1–8

[3] Tai APK, Mickley LJ and Jacob DJ 2010 Correlations between fine particulate matter (PM2.5) and meteorological variables in the United States: Implications for the sensitivity of PM2.5 to climate change *Atmos. Environ.* **44** 3976–84

[4] Yang F, Tan J, Zhao Q, Du Z, He K, Ma Y, Duan F and Chen G 2011 Characteristics of PM2.5 speciation in representative megacities and across China *Atmos. Chem. Phys.* **11** 5207-19

[5] Amil N, Latif MT, Khan MF and Mohamad M 2016 Seasonal variability of PM2.5 composition and sources in the Klang Valley urban-industrial environment *Atmos. Chem. Phys.* **16** 5357–81

[6] Draxler RR and Rolph GD 2003 HYSPLIT (HYbrid Single-Particle Lagrangian Integrated Trajectory) Model access via NOAA ARL READY Website. http://www.arl.noaa.gov/ready/hysplit4.html NOAA Air Resources Laboratory Silver Spring MD

[7] Stein AF, Draxler RR, Rolph GD, Stunder BJB, Cohen MD and Ngan F 2015 NOAA's HYSPLIT atmospheric transport and dispersion modeling system *Bull. Am. Meteorol. Soc.* **96** 2059–77

[8] Koe LCC, Arellano AF, McGregor JL 2001 Investigating the haze transport from 1997 biomass burning in Southeast Asia: its impact upon Singapore *Atmos. Environ.* **35**: 2723–34

[9] Su L, Yuan Z, Fung JC and Lau AK 2015 A Comparison of HYSPLIT backward trajectories generated from Two GDAS datasets *Sci. Total Environ.* **506–507** 527–537.

[10] Stohl A 1998 Computation, accuracy and application of trajectories - A review and bibliography *Atmos. Environ.* **32** 947–966

[11] McGowan H and Clark A 2008 Identification of dust transport pathways from lake eyre, australia using hysplit *Atmos. Environ.* **42** 6915–25

Theoretical Performance of a Multi-Cyclone Fine Particulate Emission Control System

Norelyza Hussein[1], Mohd Rashid[2], Nor Hasyimah Hashim[2], Nor Ruwaida Jamian[2], Che Hafizan Che Hassan[3]

[1] School of Civil Engineering, Faculty of Engineering, Universiti Teknologi Malaysia, 81310 Johor Bahru, Malaysia.
[2] Malaysia-Japan International Institute of Technology, UTM Kuala Lumpur, Jalan Sultan Yahya Petra, Kampung Datuk Keramat, 54100 Kuala Lumpur, Malaysia.
[3] Centre for Environmental Sustainability and Water Security (IPASA), Universiti Teknologi Malaysia,, 81310 Skudai, Johor, Malaysia

Corresponding author: norelyza@utm.my

Abstract. The study involves the development of theoretical performance background of a newly developed multi-cyclones fine particulate emission control system named as MR-deDuster. The theoretical background of the unit was developed based on the modifications of established design equations available in literatures. The Leith & Licht model and modified Lapple model with different definitions of maximum radial distance travelled by a particulate (*W*) were used to predict the performance of the unit under various volumetric gas flow rate ranging from 0.13 to 0.21 m^3/s. The result reveals significant differences among predictions, but yet all the predictions approve the capability of the unit to capture the particulate matter, especially the fine size fraction.

1. Introduction

Cyclone is one of the oldest types of particulate control equipment developed for removing or separating industrial dust from air or process gas [1-3]. Extensive studies involving experimental work on such unit started since the last decade of 19th century [4]. The data gathered from these studies was used to develop theoretical models as a simple method to predict the performance of a cyclone and many of which were based on semi-empirical ground.

Lapple (1951) introduced the first model on the prediction of the collection efficiency of a cyclone [5]. The model is based on time flight particulate approach, which is similar to the method applied in sedimentation chambers, which assumed particulate travel from the entry half width to the wall [4,6]. The time for a particulate to reach the wall is considered as gas residence time for particulate to be separated at 50% collection efficiency. This model also assumed that the particulate distribution is uniform across the entry section. The number of effective turns, N_e is fixed at average value of 5 and the tangential velocity used is assumed at constant rate. Leith & Licht (1972) introduced another model based on time flight particulate approach as well [7]. However, their model considered a continuous flux of particulate instead of single particulate applied in Lapple model. The collection efficiency equation of Leith & Licht model considered turbulent diffusion within the cyclone and the equation was derived by material balance of centrifugal force and inertia [4].

Content from this work may be used under the terms of the Creative Commons Attribution 3.0 licence. Any further distribution of this work must maintain attribution to the author(s) and the title of the work, journal citation and DOI.
Published under licence by IOP Publishing Ltd

There are many other models developed to predict the collection efficiency of cyclone, such as Dietz model (1981), Ioza & Leith model (1989), and Mothes & Loffler model (1988) [8-10]. However, the Lapple model and Leith & Licht model are the only models discussed in this study. The Lapple method is preferable due to its simplicity and it is acceptable for many situations according to Cooper and Alley (2011) [11]. Meanwhile, the Leith & Licht model is an elaborate model using the same principle of Lapple model and one of the popular models used in predicting the collection efficiency of cyclone [4].

In this paper, the theoretical prediction on the collection efficiency of a newly developed multi-cyclones using modified Lapple model, and Leith & Licht model for different volumetric gas flow rates is presented. The comparative collection efficiency between these two models was performed to highlight the capability of the unit in capturing dust, especially in the fine size fraction.

2. Methodology

The theoretical collection efficiency of MR-deDuster was predicted based on modified Lapple model, and Leith & Licht model. The existing Lapple model, as shown in equation (1), was developed based on tangential entry cyclone with the assumption of W as the width of cyclone entry, which is not as in axial entry cyclone of MR-deDuster. In this case, the W value for axial entry cyclone was assumed with different assumption from tangential entry cyclone. Three different assumptions were made to define the W value for axial entry cyclone in this study: i) W based on distance from vortex finder wall to cyclone wall, $W=D-D_e/2$, ii) W based on inlet area of cyclone entry, $W=A/0.75D$, and iii) W based on hydraulic diameter, $W= 3(D-D_e)/4$.

$$d_{pc} = \left[\frac{9\mu W}{2\pi N_e v_i(\rho_p-\rho_g)}\right]^{1/2} \tag{1}$$

where, d_{pc} is the cut diameter, μ is the gas viscosity (kg/m^3), W is the maximum radial distance of particulate (m), N_e is the number of effective turns, V_i is the gas inlet velocity (m/s), ρ_g is the density of gas (kg/m^3), and ρ_p is the density of dust (kg/m^3).

Leith & Licht model was also used to predict the collection efficiency of MR-deDuster. The Leith & Licht model used in this study is the model that has been re-arranged by Benitez (1993) for multi-cyclone unit collection efficiency prediction, as shown in equation (2) [12].

$$\eta_j = 1 - exp\left\{-2\left[\frac{kQ\tau}{MND^3}\right]^{\frac{M}{2}}\right\} \tag{2}$$

where, η_j is the collection efficiency, Q is the volumetric gas flow rate, k is a dimensionless geometric configuration parameter, N is the number of miniature cyclones installed in the multi-cyclones, D is a diameter of single miniature cyclone, whereas M and τ are described coefficient related to cyclone configuration, temperature, pressure, and velocity of gas flow.

The performance prediction was based on four units of MR-deDuster arranged as a multicycone system. The performance prediction of the unit was studied under four different volumetric air flow rates of 0.13, 0.16, 0.19, and 0.21 m^3/s with respective inlet velocity of 13, 16, 19, and 21 m/s.

3. Results and Discussions

3.1. Theoretical Collection Efficiency

Table 1 shows the cut diameter prediction of MR-deDuster for different models at various volumetric gas flow rates of 0.13, 0.16, 0.19, and 0.21 m^3/s, which depicted that the cut diameter reduces as the volumetric gas flow rate increases and the trend was applied to all models tested. The smallest cut diameter for all models is obtained at volumetric gas flow rate of 0.21 m^3/s, while figure 1 depicts the fractional collection efficiency prediction of the unit based on the different theoretical models for various volumetric gas flow rates, which showed that the fractional collection efficiency increases with the increasing of volumetric gas flow rate. The highest fractional collection efficiency of each particulate size fraction was obtained at the highest volumetric gas flow rate of 0.21 m^3/s. The overall collection efficiency of all models tested is shown in table 2. The overall collection efficiency of the unit showed similar trend as fractional collection efficiency, which indicates that the overall collection efficiency increases with the increasing of volumetric gas flow rate.

Table 1. Cut diameter prediction of MR-deDuster for different models at various volumetric gas flow rates

Volumetric Flow Rate (Q), m^3/s	Cut Diameter (d_{pc}), µm			
	Lapple: $W = D\text{-}D_e/2$	Lapple: $W = A/\ 0.75D$	Lapple: $W = 3(D\text{-}D_e)/4$	Leith & Licht
0.13	1.8	3.6	2.2	2.5
0.16	1.7	3.2	2.0	2.2
0.19	1.5	3.0	1.8	1.9
0.21	1.4	2.8	1.7	1.7

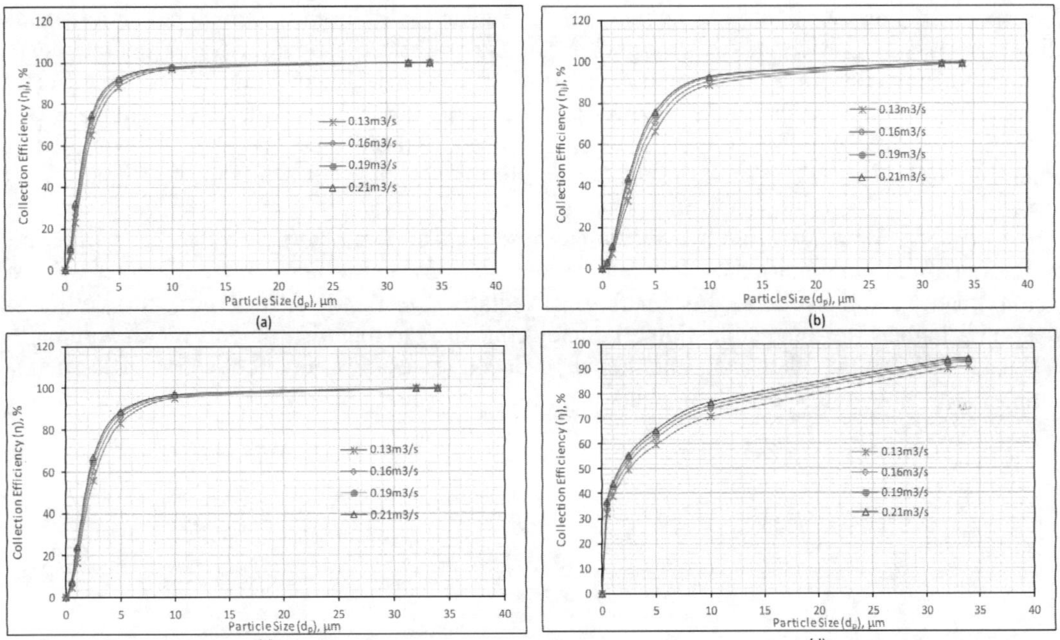

Figure 1. The fractional collection efficiency prediction of MR-deDuster for different models (a) Lapple model: $W = D\text{-}D_e/2$ (b) Lapple model: $W = A/0.75D$ (c) Lapple model: $W = 3(D\text{-}D_e)/4$ (d) Leith & Licht model

Table 2. Overall collection efficiency prediction of MR-deDuster for different models at various volumetric gas flow rates

Volumetric Flow Rate (Q), m^3/s	Overall Collection Efficiency (η_o), %			
	Lapple: $W = D\text{-}D_e/2$	Lapple: $W = A/0.75D$	Lapple: $W = 3(D\text{-}D_e)/4$	Leith & Licht
0.13	96.90	92.92	96.02	84.79
0.16	97.30	93.73	96.51	86.49
0.19	97.60	94.34	96.88	87.81
0.21	97.75	94.67	97.08	88.53

Figure 2 (a-d) presents a comparison between fractional collection efficiency predictions of MR-deDuster based on different theoretical models for volumetric gas flow rates of 0.13, 0.16, 0.19, and 0.21 m^3/s. As can be observed from the figures, the fractional collection efficiency predictions differ widely, especially for fine size fraction particulate. Consequently, the choice of model significantly affects the predicted PM_{10} and, especially, $PM_{2.5}$ efficiencies. For example, the predicted $PM_{2.5}$ efficiency of cyclone operating at volumetric gas flow rate of 0.13 m^3/s can be higher than 60%, according to Lapple: $W=D\text{-}D_e/2$, and lower than 35%, according to Lapple: $W=A/0.75D$.

The model of Leith & Licht yields exponential type curve with high fractional collection efficiency for fine size fraction particulate (i.e., more than 50% of collection efficiency for particulate is smaller and equals to 1μm in size for all volumetric gas flow rates tested). However, the collection efficiency of Leith & Licht model was increased slower compared to other models as the particulate coarser. Meanwhile, the three modified Lapple models produced sigmoid type of curve. Among all models, Lapple: $W=D\text{-}D_e/2$ and Lapple: $W= 3(D\text{-}D_e)/4$ indicate yield curves with fast fractional collection efficiency transient from low values to high values.

The model of Lapple: $W=D\text{-}D_e/2$, yields higher fractional collection efficiency for most of particulate size fractions for all volumetric gas flow rates tested. This is due to the small cut diameter value of the model compared to other models. As for fine particulates size fractions (d_p<1μm), the Leith & Licht model yields higher fractional collection efficiency compared to the other models for all volumetric gas flow rates tested. On the other hand, the model produced lowest fractional collection efficiency for all volumetric gas flow rates tested for coarser particulates (d_p>5μm), which later affects the overall collection efficiency.

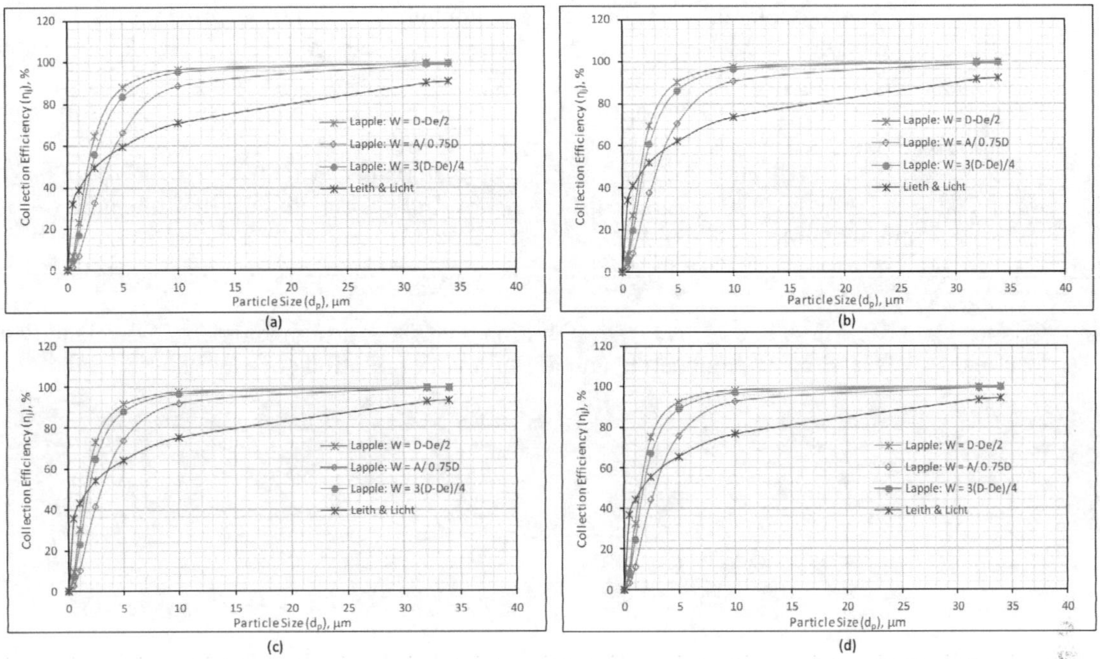

Figure 2. Comparison of fractional collection efficiency prediction models of MR-deDuster for different volumetric gas flow rates (a) 0.13m^3/s (b) 0.16m^3/s (c) 0.19m^3/s (d) 0.21m^3/s

4. Conclusion

The study predicted that the increasing of volumetric gas flow rate results in a reduction of cut diameter, and an increase of fractional collection efficiency, and overall collection efficiency. These trends are applied to all theoretical models used in this study. Even all tested models show similar trends, the results show that the collection efficiency predictions differ significantly, especially for fine size fraction particulate. Consequently, the choice of model significantly affects the predicted PM_{10} and, especially, $PM_{2.5}$ efficiencies. The study indicated that the Lapple model: *W=D-D$_e$/2* exhibits the smallest cut diameter and the highest overall collection efficiency, while Lapple model: *W=A/0.75D* displays the largest cut diameter and low overall collection efficiency. However, Leith & Licht model exhibits the lowest overall collection efficiency despite of having small cut diameter and high sensitivity towards volumetric gas flow rate changes. This is due to lower fractional collection efficiency of coarse particulate, which represents most mass percentage of overall mass fraction. The findings illustrate that the overall collection efficiency not only depend on cut diameter, but also depends on the fractional collection efficiency of each size fraction.

References

[1] Rashid, M., Huda, N., Norelyza, H., & Hasyimah, N. (2015). Comparison of the Performance of MR-deDuster with Other Conventional Cyclones. Sains Malaysiana, 44(4), 565-569.

[2] Norelyza, H., Rashid, M., Hajar, S., & Nurnadia, A. (2014). MR-deDuster: A dust emission separator in air pollution control. Jurnal Teknologi, 68(5).

[3] Norelyza, H., & Rashid, M. (2013). Performance of MR-deDuster: A case study of a palm oil mill plant. In Advanced Materials Research (Vol. 664, pp. 133-137). Trans Tech Publications.

[4] Cortes, C., & Gil, A. (2007). Modeling the gas and particle flow inside cyclone separators. Progress in Energy and Combustion Science, 33(5), 409-452.

[5] Lapple, C. E. (1951). Processes use many collector types. Chemical Engineering, 58(5), 144-151.

[6] Marinos, C., Economopoulou, A. A., & Economopoulos, A. P. (2007). Comparative fractional efficiency predictions by selected cyclone simulation models. Journal of Environmental Engineering, 133(5), 556-567.
[7] Leith, D., & Licht, W. (1972). The collection efficiency of cyclone type particle collectors: a new theoretical approach. AIChE Symp. Ser. 68(126), 196-206.
[8] Dietz, P. W. (1981). Collection efficiency of cyclone separators. AIChE Journal, 27(6), 888-892.
[9] Iozia, D. L., & Leith, D. (1990). The logistic function and cyclone fractional efficiency. Aerosol Science and Technology, 12(3), 598-606.
[10] Mothes, H., & Löffler, F. (1988). Prediction of particle removal in cyclone separators. Int. Chem. Eng, 28(2), 231-240.
[11] Cooper, C. D., & Alley, F. C. (2011). Air pollution control: A design approach. Waveland Press.
[12] Benitez, J. (1993). Process engineering and design for air pollution control. Prentice Hall.

IOP Conf. Series: Materials Science and Engineering **636** (2019) 012007 doi:10.1088/1757-899X/636/1/012007

Effectiveness of Local Waste Materials as Organic-Based Coagulant in Treating Water

N S Zaidi[1*], K Muda[1], M A Abdul Rahman[1], M S Sgawi[1] and A H Amran[1]
*Corresponding author: +60177685310 (Tel. No.); +6075566157 (Fax No.); nursyamimi@utm.my (Email)
[1] School of Civil Engineering, Faculty of Engineering, Universiti Teknologi Malaysia, 81310 UTM Johor Bahru, Johor, Malaysia.

Abstract. The adequacy of chemicals as coagulants such as alum is all around perceived. Nevertheless, there are numerous drawbacks related with the use including high operational costs, impeding impacts on human wellbeing and generation of huge sludge volumes. In this manner, it is crucial to supplant these chemical coagulants with organic-based coagulants to minimize the downsides. Therefore, the aim of this study is to identify the suitability of selected local wastes in becoming organic-based coagulant as to treat raw water. Regarding choice of organic-based coagulants, this study focused on the local waste materials, which are chestnut peels, bagasse and maize cobs. These waste materials were prepared by washing, drying, grinding lastly sieving in obtaining powder of coagulants ready to be used. A jar test was then performed to decide the effect of individual characteristic coagulants on the effectiveness of turbidity removal under different working variables of pH and coagulant dosage. Based on the findings, bagasse contributed to the highest yield of 79.5% followed by chestnut and maize cob. From the jar test experiments, the optimum dosage of 90 mg/L and pH 7.5 was obtained. At these optimizations, highest turbidity removal of 97.3% was recorded by bagasse compared to the other tested natural coagulants. This showing that bagasse as one of the representatives of organic wastes has the potential to physically treat the water.
Keywords: Organic-based coagulant; Organic wastes; Coagulant dosage; Turbidity removal.

1. Introduction

In conventional water and wastewater treatment, coagulation and flocculation is one of the important processes [8]. Coagulation is a process whereby the particles adhered within each other thus, forming larger flocs which subsequently can easily settle at later stage. The addition of coagulant helps to destabilize the suspended colloidal particles hence, promoting subsequent coagulation [14]. In water, there is presence of colloidal particles which posing same negative charges. By adding coagulant, the neutralization process of these charges can be occurred.

Coagulation can be described as a complex chemical reaction that makes its dosing addition exhibits non-linear behaviour [2]. This makes the control of coagulation process in water and wastewater treatment is quite difficult. Uniform dispersion of coagulant in water or wastewater is important to increase particle-to-particle contact. Subsequent gentle and prolonged mixing times are among factors that are required to ensure uniform dispersion of coagulant. This is to encourage agglomeration of microscopic particles into larger flocs. Suspended polluting matter will adsorb to the large flocs. With more adsorption, the coagulated particles will increase in size and mass and finally will settle to the bottom of the tank once it is saturated with adsorbed suspended pollutants [12].

Until today, the conventional coagulant used is a chemical-based coagulant. There are various types of chemical coagulants such as aluminum sulfate (alum), ferric chloride, cationic polymers and

Content from this work may be used under the terms of the Creative Commons Attribution 3.0 licence. Any further distribution of this work must maintain attribution to the author(s) and the title of the work, journal citation and DOI.
Published under licence by IOP Publishing Ltd

many others [1, 4, 6, 10]. Currently, in the water treatment process, compound that mostly used as a coagulant is the alum. Alum is capable in exhibiting good results for turbidity and color removal of raw water. In any case, there are numerous downsides related with the utilization of these coagulants. For example, high operational costs, inconvenient consequences for human wellbeing and the way that it is essentially influence pH of treated water. The utilization of chemical coagulants likewise will generally produce more end sludge product and for the most part, these sludge are harmful and danger. Due to such issue, legitimate treatment of the sludge before transfer is imperative. This sludge will be disposed at Secured Landfill, which at that point require more prominent expense [7].

The option of chemical coagulant, which is organic-based coagulant or also known as natural coagulant has been utilized in many countries as an effective coagulant. These coagulants were produced from different materials such as plant seeds, leaves and roots. Due to its promising results, a lot of developing countries are now using natural coagulant in their water and wastewater treatment on small scale. Among the common and established natural coagulants are *Moringa oleifera*, *Jatropha curcas*, *Strychnos potatorum*, and *Hibiscus sabdariffa* [11]. These coagulants are sustainable as there are less wellbeing risks and the expense of delivering these natural coagulants were more affordable than the chemical coagulants since it is locally accessible. The natural coagulants can also treat water of high turbidity and having remarkable removal proficiency [3, 5].

To date, the chemical coagulants are ending up less sustainable for coagulation of water and wastewater treatment due to different adverse impacts. It is currently crucial to supplant these coagulants with organic-based coagulants that can minimize the previously mentioned downsides. Therefore, the objective of this study is to identify the suitability of a selected local wastes in becoming natural coagulant as to treat raw turbid water. The feasibility analysis will look into the coagulant's physical characteristics and its potential in removing turbidity of a water.

2. Materials and Methods

2.1 Experimental procedure

In this study, raw water sample was collected from Water Treatment Plant, Taman Impian Emas, Skudai. The reason of this choice is that the water from this plant is a source of water supply to the consumers around Skudai area. The collected water sample has initial turbidity of 100 to 200 NTU. This water sample was treated using prepared organic-based coagulant and its effectiveness in terms of turbidity removal and coagulation activity is identified. For the preparation of natural coagulant, the chosen fruit seeds or fruit peels undergoes several preparation stages involving washing, drying, grinding, and sieving. These processes are conducted until final product in form of powder is obtained. The selected plant-based materials were bagasse, whole chestnut, maize cob and chestnut peel. The reason of these choices is that such materials are less likely been investigated until now. These materials were collected from the local markets and neighbourhoods. The prepared coagulant is then characterized for dry weight, yield and moisture content.

The study is the continued to determine the effect of pH and coagulant's dosage upon the efficiency of organic-based natural coagulants in reducing the turbidity. When the pH is constant at 7.5, the dosages were varied at 10 mg/L, 30 mg/L, 50 mg/L, 70 mg/L, 90 mg/L and 110 mg/L. This experiment resulted in an optimum dosage to be used at later experiment. Consecutively, the pH was varied at pH 2, 4, 7.5, 10 and 12 under the optimized dosage that obtained previously. In this study, sodium hydroxide (NaOH) of 2.0M was used to enhance the coagulant agent. The optimization of best dosage and pH for each of the prepared natural coagulants was determined based on the highest coagulation activity and turbidity removal.

2.2 Analytical methods

The prepared organic-based coagulants are characterized for yield and moisture contents. Yield is amount of the dry powder from the initial raw material of the selected organic-based natural coagulant. As for moisture content, it was analysed in order to determine how much water content in each prepared coagulant. The moisture content can be expressed in percentage as shown in Eq. (1).

$$\text{Moisture content (\%)} = \frac{(W_o - W_e)}{W_o} \times 100 \quad (1)$$

where,
W_o: Weight before drying (g)
W_e: Weight after drying (powder) (g)

Turbidity analysis was conducted by first preparing 500 mL of raw water sample prior and after the addition of coagulant of specified dosage. The sample was then placed in a glass cuvette of 10 mL and then measured using 2100P (HACH) Portable Turbidimeter. In order to investigate the potential of coagulation for the particles in turbid water under the effect of various natural coagulants, the parameter of coagulation activity is then conducted. The final turbidity of water sample after treated with respective natural coagulant of specified dosage (F_e) was taken during Jar Test. Additionally, the final turbidity of blank sample (without addition of natural coagulant) was as well recorded (F_o). The percentage of coagulation activity can be calculated using Eq. (2).

$$\text{Coagulation activity (\%)} = \frac{(F_o - F_e)}{F_o} \times 100 \quad (2)$$

where,
F_o: Final turbidity for blank water sample (NTU)
F_e: Final turbidity for natural coagulant (NTU)

3. Results and Discussions

3.1 Yield of the selected organic-based coagulant

Based on Figure 1, whole chestnut contributed to the highest yield with initial weight recorded as only 265g with resulted yield was 170g. This contributes to 64.2% of the whole prepared material in becoming an accessible organic-based coagulant. The second highest yield was contributed by maize cob. This particular material shows major difference of weight between initial weight, which was 345g and yield of 80g. This contributes to 23.2% of the whole prepared material in becoming the accessible organic-based coagulant. The percentage is quite low, thus showing that maize cob is a least preferable material to be chosen because the preparation of the initial material needs to be abundant with the final production of the coagulant is very least. Meanwhile, bagasse and chestnut peel resulted in a quite similar yield. Based on Figure 1, it can be seen that the weight of initial material for bagasse is 88g with its yield recorded as 70g. This contributes to 79.5% of the whole prepared material in becoming the accessible organic-based coagulant.

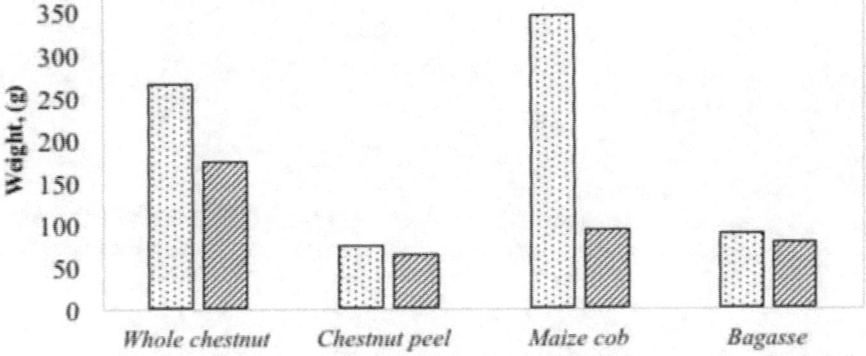

Figure 1. Experimental results of yield (Initial materials; Yield)

IOP Conf. Series: Materials Science and Engineering **636** (2019) 012007 doi:10.1088/1757-899X/636/1/012007

As for chestnut peel, the weight of initial material recorded as 74g with its yield of 58g thus, contributes to 78.4% of the whole prepared material in becoming the accessible organic-based coagulant. Highest percentage indicating that the final form of coagulants can be easily generated from the small batch of initially prepared materials. Overall, the findings concluded that bagasse and chestnut peel are among the feasible materials to work with in producing the final form of organic-based coagulant. This followed by whole chestnut and lastly, maize cob. Nevertheless, in terms of preparation of materials at initial stage, whole chestnut seemed to be easy to gather as greater weight can easily be obtained compared to bagasse and chestnut peel.

Table 1 shows the dry weight and moisture content for each of the prepared natural coagulants.Based on Table 1, the measured values of moisture content reveals variability with the range between 13 to 75%. The moisture content for maize cob is higher with 74.5% compared to the other natural coagulants. This is because the water content in the maize cobs evaporated quite slowduringthe prior drying process. Such high moisture contents resulted to a difficulty to obtain yield at the end of preparation process. This results supported the observation of yield for maize cob that is lesser than the other materials.

Table 1. Dry weight and moisture contents of the prepared organic-based coagulants

Natural Coagulants	Initial Weight (g)	Dry Weight (g)	Moisture Content (%)
Whole chestnut	265	173	34.7
Chestnut peel	74	61	17.6
Maize cob	345	88	74.5
Bagasse	88	76	13.6

On the other hand, chestnut peel and bagasse were both contained low moisture content of 17.6% and 13.6%, respectively. Too low moisture content is also not very good as most of the coagulant agent in the material will be dried, therefore might not be effective during coagulation process. According to Madukwe *et al.* [9], the moisture content for dry *Moringa oleifera* leaf powder was 7.1% which is very low and not suitable to become alternative for chemical based coagulant in water treatment. Meanwhile, Shilpa *et al.* [13] also reported that *Hyacinth* bean peels has lower moisture content of about 9.8% thus, showed less ability to be one of the best natural coagulant. The remaining material, which is whole chestnut containing a moderate moisture content of 34.7%, thus resulted in greater yield among all other natural coagulants.

3.2 Effect of dosage and pH of selected natural coagulant in treating water

As for coagulant's dosage and pH, Figure 2 and Figure 3 represents the profile of turbidity removal when various dosages and pH were experimented.

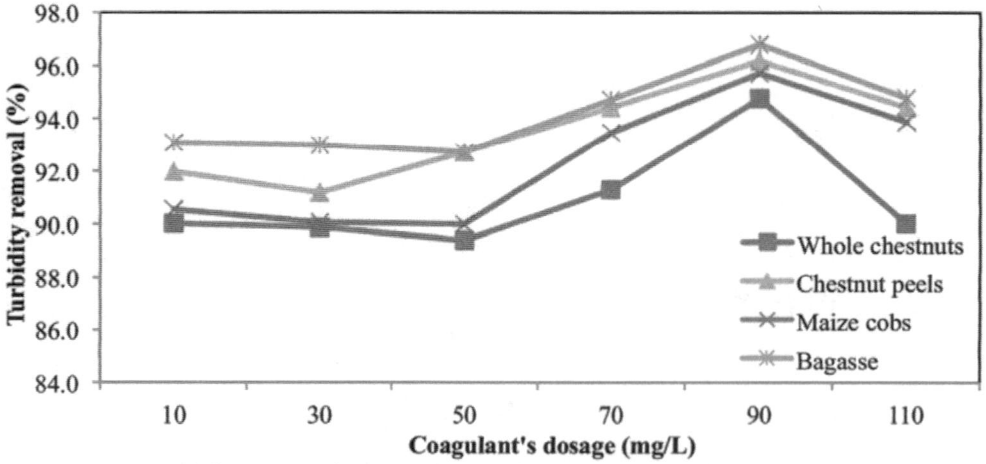

Figure 2. Turbidity removal of tested natural coagulants at different dosages (at pH 7.5)

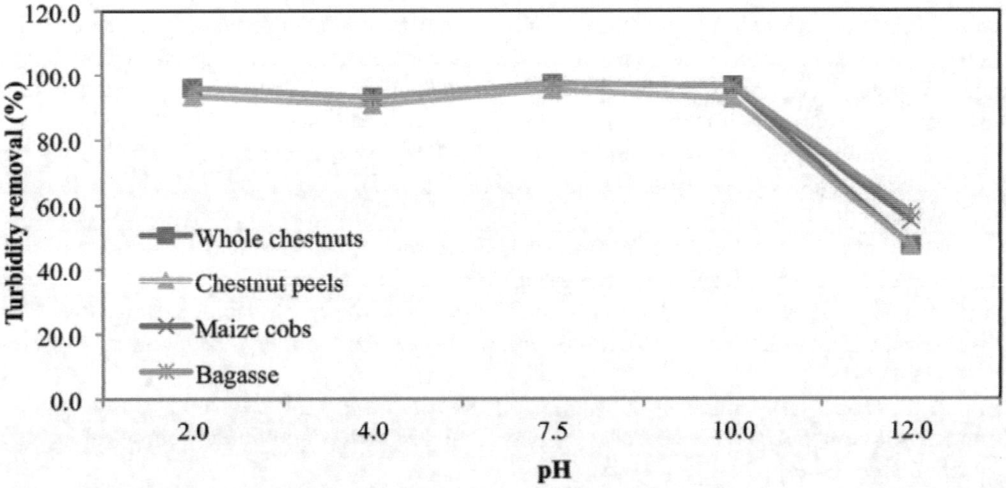

Figure 3. Turbidity removal of tested natural coagulants at different pH (at dosage 90 mg/L)

Based on Figure 2, it can be seen that the optimum dosage is 90 mg/L. Bagasse is recorded as the natural coagulant that resulted in higher percentage of turbidity removal of approximate 96.0%, followed by chestnut peel, maize cob and whole chestnut. Although, all natural coagulants resulted in almost similar range of turbidity removal at 90 mg/L, the starting percentage is rather high for bagasse compared to the other coagulants. Hence, makes bagasse as a feasible natural coagulant to be used in the treatment. Meanwhile, based on Figure 3, it can be seen that the optimum pH is 7.5. Yet again, bagasse is recorded as the natural coagulant that resulted in higher percentage of turbidity removal by 97.3%, followed by maize cob, whole chestnut and chestnut peel. Although all natural coagulants resulted in almost similar range of turbidity removal at pH 7.5, the starting percentage is rather high for bagasse compared to the other tested coagulants. This suggesting that bagasse is a feasible natural coagulant to be used in treating polluted river water.

Conclusions

Local waste materials are potential organic-based coagulants that can be used to treat water sustainably. On top of its significant effect on treatment removal performances, organic-based natural coagulants have also assembled great interests worldwide due to its ease of biodegradability and environmental friendly nature. Main contribution of this study focuses on the perspective that the organic-based coagulant can pose remarkable treatment performance, at par with chemical coagulant. Moreover, the uses of organic-based coagulants were all based from local plant wastes, which can help in reducing the amount of wastes to be disposed at sanitary landfill. Overall, the conclusions that can be drawn from this study are as the following below.

1. Bagasse is among the feasible materials to work with in producing the final form of organic-based coagulant, followed by chestnut and maize cob. However, in terms of preparation of materials at initial stage, whole chestnut is easy to gather as greater weight can be easily obtained.
2. Based on Jar Test experiments, the optimum dosage of 90 mg/L and pH 7.5 was obtained. At these optimizations, highest turbidity removal of 97.3% was recorded by bagasse compared to the other tested organic-based coagulants.

Acknowledgement

The authors thank the Universiti Teknologi Malaysia (UTM) for financially supported this research using Research University Grant (RUG) (Grant No.: 14J41).

IOP Conf. Series: Materials Science and Engineering **636** (2019) 012007 doi:10.1088/1757-899X/636/1/012007

References

[1] Aboulhassan M A, Souabi S, Yaacoubi A and Baudu M 2006. Improvement of paint effluents coagulation using natural and synthetic aids. *Journal of Hazardous Materials*. **138** 40–45.

[2] Adgar A, Cox C S and Jones C A 2005. Enhancement of coagulation control using the streaming current detector. *Bioprocess and Biosystems Engineering*. **27** 349–357.

[3] Amran A H, Zaidi N S, Muda K and Loan L W 2018. Effectiveness of Natural Coagulant in Coagulation Process: A Review. *International Journal of Engineering & Technology*. **7(3.9)** 34-37.

[4] Baptista A T A, Coldebella P F, Cardines P H F, Gomes R G, Vieira M F, Bergamasco R and Vieira A M S 2015. *Chemical Engineering Journal*. **276** 166-173.

[5] Bhuptawat H, Folkard G K and Chaudhari S 2007. Innovative physico-chemical treatment of wastewater incorporating *Moringa oleifera* seed coagulant. *Journal of Hazardous Materials*. **142(1)** 477-482.

[6] Chang Y S, Kim J H, Murugesan K, Kim Y M, Kim E J and Jeon J R 2009. Use of grape seed and its natural polyphenol extracts as a natural organic coagulant for removal of cationic dyes. *Chemosphere*. **77** 1090–1098.

[7] Huang X, Bo X, Zhao Y, Gao B, Wang Y, Sun S, Yue Q and Li Q 2014. Effects of compound bioflocculant on coagulation performance and floc properties for dye removal. *Bioresource Technology*. **165** 116–121.

[8] Jiang J Q 2015. The role of coagulation in water treatment. *Current Opinion in Chemical Engineering*. **8** 36-44.

[9] Madukwe E U, Ugwuoke A L and Ezeugwu J O 2013. Effectiveness of dry *Moringa oleifera* leaf powder in treatment of anaemia. *International Journal of Medicine and Medical Sciences*. **5(5)** 226-228.

[10] Okuda T, Baes A U, Nishijima W and Okada M 1999. Improvement of extraction method of coagulation active components from *Moringa oleifera* seed. *Water Research*. **33(15)** 3373–3378.

[11] Pritchard M, Mkandawire T, Edmondson A, O'neill J G and Kululanga G 2009. Potential of using plant extracts for purification of shallow well water in Malawi. *Physics and Chemistry of the Earth, Parts A/B/C*. **34(13-16)** 799-805.

[12] Ratnaweera H and Fettig J 2015. State of the art of online monitoring and control of the coagulation process. *Water*. **7(11)** 6574-6597.

[13] Shilpa B S, Akanksha K and Girish P 2012. Evaluation of cactus and hyacinth bean peels as natural coagulants. *International Journal of Chemistry and Environmental Engineering*. **3(3)** 187–191.

[14] Zueva S B, Ostrikov A N, Ilyina N M, De Michelis I and Vegliò F (2013). Coagulation processes for treatment of wastewater from meat industry. *International Journal of Waste Resources*. **3** 130.

SERVICE FREQUENCY AND SERVICE HOURS EVALUATION FOR BUS SERVICE IN WEST KLANG VALLEY

Shuhairy Norhisham[1, 2], Muhammad Fadhlullah Abu Bakar[1], Zainul Hazwani Abd Samad[1], Agusril Syamsir[1,2], Nur'atiah Zaini[1], Herda Yati Katman[1], Mohd Zakwan Ramli[1]

[1]College of Engineering, Universiti Tenaga Nasional, 43000 Kajang, Selangor, Malaysia.
[2] Institute Of Energy Infrastructures (IEI), Universiti Tenaga Nasional, 43000 Kajang, Selangor, Malaysia.
*Corresponding author's email: Shuhairy@uniten.edu.my

Abstract. The Malaysian government has taken an initiative to mitigate public transport issues in the Klang Valley through the Land Public Transport Commission (SPAD). Several issues were raised by passengers in Klang Valley in regard to public transport, such as long waiting time and poor management of bus service hours. This study is aimed to investigate the service frequency for bus service performance and the public bus service hours in Klang Valley. Five different areas were selected, which consisted of three routes for each area in west Klang Valley, which were Klang, Shah Alam, Subang Jaya, Petaling Jaya, and Puchong. The time interval for every scheduled bus departure, average time interval, and bus service hours were determined based on the collected data. The result showed that Puchong has the shortest average time interval for service quality. Therefore, the bus performance with regard to service frequency and service hours should be improved to provide the public with better service.

1. Introduction

In 2010, less than 20% or approximately 1.24 million daily trips in Klang Valley used the public transport [3]. RapidKL is the major bus company that operates in urban areas of Klang Valley. It is a service brand operated by Prasarana Malaysia. The Klang Valley Network Revamp was established by the Land Public Transport Commission (SPAD) in December 2015 [2]. Other bus operators include Metrobus, Selangor Omnibus, Len Seng, Transnasional/*Kenderaan* Klang-Banting, Triton, and Permata Kiara, which are operated by Intrakota and Cityliner [4]. Another new addition to the Klang Valley bus services are the 120 RapidKL buses, which include 40 double-deckers that hit the Klang Valley streets in September. These RapidKL buses are capable of picking up an additional of 10,000 passengers a day when they operatinalise .The buses, especially the double-deckers, are expected to service the region's most heavily-used roads [7, 5].

All public transport modes, have several attributes that will affect the service performance. Bus service performance can be defined by several attributes, such as service hour, service frequency, and passenger load factor [11, 20]. These common attributes will determine factors on comfort and convenience that contribute to passenger satisfaction with the service and their likelihood to use it. Another factor that contributes to the bus frequency issue is the unpredictable traffic condition in the Klang Valley [4, 9]. It could be a major factor that cause a delay in service frequency and time travel to arrive at a selected distance. In this context, RapidKL have suggested to the authorities to provide bus lanes to show its support for public transport service. The unpredictable movements of public transport modes have caused difficulty for the public to plan their journeys effectively [10, 12].

Content from this work may be used under the terms of the Creative Commons Attribution 3.0 licence. Any further distribution of this work must maintain attribution to the author(s) and the title of the work, journal citation and DOI.
Published under licence by IOP Publishing Ltd

In addition, passengers in the Klang Valley are plagued with numerous problems, such as long waiting time (service frequency), overcrowding, especially during peak hours (passenger load factor), and poor management of bus service hours, which will cause delays (service hour) [1, 14]. The issues should be able to identify why bus services in Klang Valley have poor performance. On the other hand, the service quality, particularly in service frequency [14], service hours, and passenger load factors, is at a satisfactory level [15]. Low frequencies also cause difficulty to fit transit trips into the activity time schedules that are carried out at different locations, and thus, may be an important impediment to the use of public transport [6, 8].

The hour or time of study is vital as the most crucial or peak hours of the day should be chosen to obtain the most critical results as possible in order to determine the quality of service (QOS) of a transit system [11]. A research conducted by Universiti Sains Malaysia on vehicles at signalised intersections was reviewed. Furthermore, the data collection time period was divided into three peak hours per day, which were morning (7.30am - 9.30am), afternoon (12.00pm-2.00pm), and evening (4.30pm-6.30pm) [19]. The real-time bus information for a particular area is provided by a bus service application that can be downloaded onto a smartphone. Therefore, it is much easier as identified the real-time bus information and its location as well [16]. In this study the service frequency for bus service performance and the service hours of public bus will be evaluated.

2. Methodology

Data collection was conducted at selected RapidKL bus routes. The total capacity for RapidKL bus is 65 passengers, who consists of 25 comfortable sitting and 40 standing passengers. A bus schedule timetable was used to determine the service frequency and evaluate the average time for bus departure time interval. The bus schedule timetable was obtained from RapidKL bus counter at the station where the route began. Five different areas were selected, which consisted of 3 routes for each area in west Klang Valley. The routes were Klang (Routes KLG1, KLG2 and 704), Shah Alam (Routes T757, T758 and SA01), Subang Jaya (Routes T776, T777 and T778), Petaling Jaya (Routes PJ02, PJ03 and PJ04), and Puchong (Routes T600, T604 and T605). These areas have been selected as study areas because these cities could be considered as major cities in West Klang Valley. All the necessary information regarding schedule of bus service in these study areas could be obtained from bus counter at the station where the route begins. Table 1 shows the QOS for service frequency and classifies the quality of service based on average departure time interval.

Table 1: Quality of service (QOS) for service frequency
(Sources: *Transit Capacity and Quality of Service Manual* [18]).

Quality of Service	Average Departure Interval (min)	Comments
A	<10	Passengers do not require schedules
B	10 – 14	Frequent services, passengers consult schedules
C	15 – 20	Maximum desirable time to wait if a bus is missed
D	21 – 30	Service unattractive to choice riders
E	31 – 60	Service available during the hours
F	>60	Service unattractive to all riders

The bus service hours for a particular route can be obtained from the service operator and it will be considered from the first service trip until the last service trip. It can also be obtained from the website, which is updated by the Land Public Transport Commission (SPAD) for Puchong (Route T600, T604, T605, T601, T602, T603, SJ03), Petaling Jaya (Route PJ02, PJ03, PJ04, Citra Mall,

IOP Conf. Series: Materials Science and Engineering **636** (2019) 012008 doi:10.1088/1757-899X/636/1/012008

Paradigm Mall, PJ01, T780, T781, T782, T783, T784, T785, T786, T787, 802, T800), Subang Jaya (Route T776, T777, T778, SJ01), Shah Alam (Route T757, T758, SA01, T771, 754, T750, T751, T752, T753, T754, T756, SA03, SA04, SA05), and Klang (Route KLG1, KLG2, 704, T710, T712, T709, T713, T700, T701, T702, T703, T707), respectively, and service hours were recorded. Hence, by comparing the service hours between service operators, the service performance quality can be obtained, as shown in Table 2. The results are presented in a graph to compare all QOSs in every route and a QOS summary for the 3 routes in each five areas was concluded.

Table 2: Quality of services (QOS) for service hours
(Sources: Transit capacity and quality of service manual [18]).

Quality of Services	Service Hours	Comments
A	19 - 24	Night or "owl" service provided
B	17 - 18	Late evening service provided
C	14 - 16	Early evening service provided
D	12 - 13	Daytime service provided
E	4 - 11	Only peak hour service or limited mid-day service
F	0 - 3	Very limited or no service

3. Result and Analysis

The data for service frequency were collected for each selected route in each location. All data are summarised in Figure 1. The three routes, which are from Klang (Route KLG1, KLG2 and 704), Shah Alam (Route T757, T758 and SA01), Subang Jaya (Route T776, T777 and T778), Petaling Jaya (Route PJ02, PJ03 and PJ04), and Puchong (Route T600, T604 and T605), were presented as RouteA, Route B, and Route C, respectively.

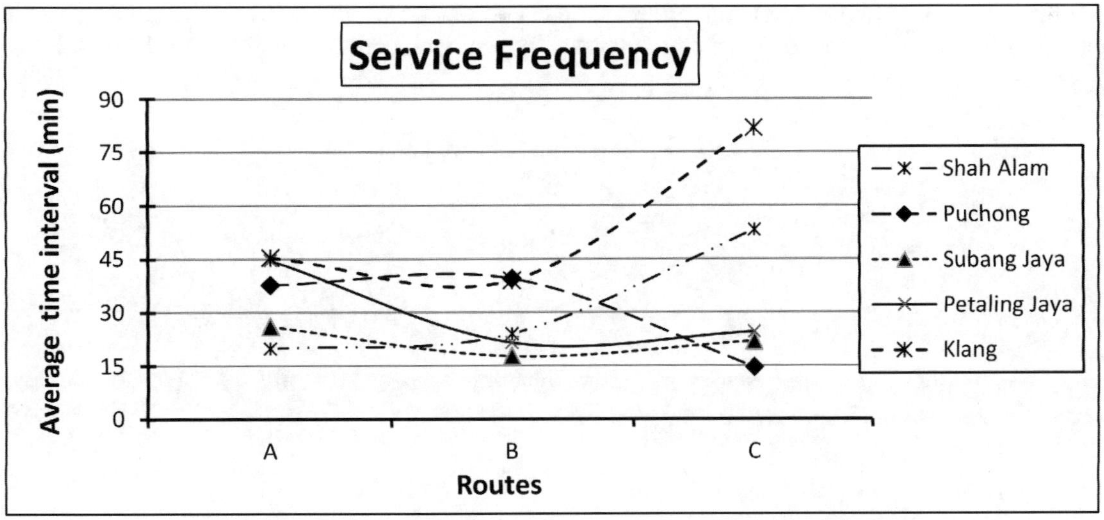

Figure 1: The service frequency graph at selected locations

As routes for those five areas did not have fixed service frequencies, the numbers of intervals were determined according to minutes per interval. The average interval for service frequency at Puchong and Subang Jaya, whose three routes were served by RapidKL, fell into QOS D. According to Transit Capacity and Quality of Service Manual, the average departure intervals for QOS D were in the range of 21-30 minutes, and were unattractive to all riders. The average departure interval for Petaling Jaya also fell into QOS D, but it is a free bus service initiated by Majlis Bandaraya Petaling Jaya (MBPJ) and served by PJ City Bus. On the other hand, the average intervals for Shah Alam fell into QOS E, which is served by RapidKL for Route T757 (A) and Route T758 (B), while Route SA01(C) is served by SMART Selangor bus. As for QOS E, the average departure interval was in the range of 31-60 minutes, where the service was available during the hours. In addition, the same was with Klang, which also fell into QOS E, but is served by SMART Selangor for Route KLG1 (A) and Route KLG2 (B), while Route 704(C) is served by Seranas. Therefore, the overall QOS, as in average for the five areas in West Klang Valley, was QOS D.

According to the manual, a QOS D gives an average headway of 16 - 30 minutes. From the passenger's perspective, passengers will have to check the scheduled arrival times to reduce waiting time. Furthermore, passengers are also required to adapt to the service schedule, which will in turn affect their usual arrival and departure times. But somehow in this case, fixed scheduled times were not provided by all operators. Passengers did not know at what time the next bus was scheduled to arrive. Apart from that, passengers could not measure their arrival and departure times. The arrivals and departures depended solely on the bus driver and the traffic flow condition at that period of time; hence, there was weakness and disadvantage in the system based on only the estimated range in service frequency. Besides, from the operator's perspective, to sustain the service frequency, two or three buses should be provided in an hour to achieve a20-30 minute headway to sustain the service frequency. However, if the traffic condition gets worse, the operator may not have the capacity to increase service frequencies.

The data for service hours were collected for each selected route at each location. Based on the data provided by the Land Public Transport Commission (SPAD) website for Shah Alam, Puchong, Subang Jaya, Petaling Jaya, and Klang, the location consisted of 14, 7, 4, 16, and 12 routes, respectively. The results for bus service hours are summarised in Figure 2. Service hour prescribes the number of operating hours for a particular bus route on a given day. In regard to the SPAD website information, the operating hours for RapidKL buses are varied and depend on the traffic flow condition or the number of buses that operate on thats particular day, which may increase, maintain, or decrease.

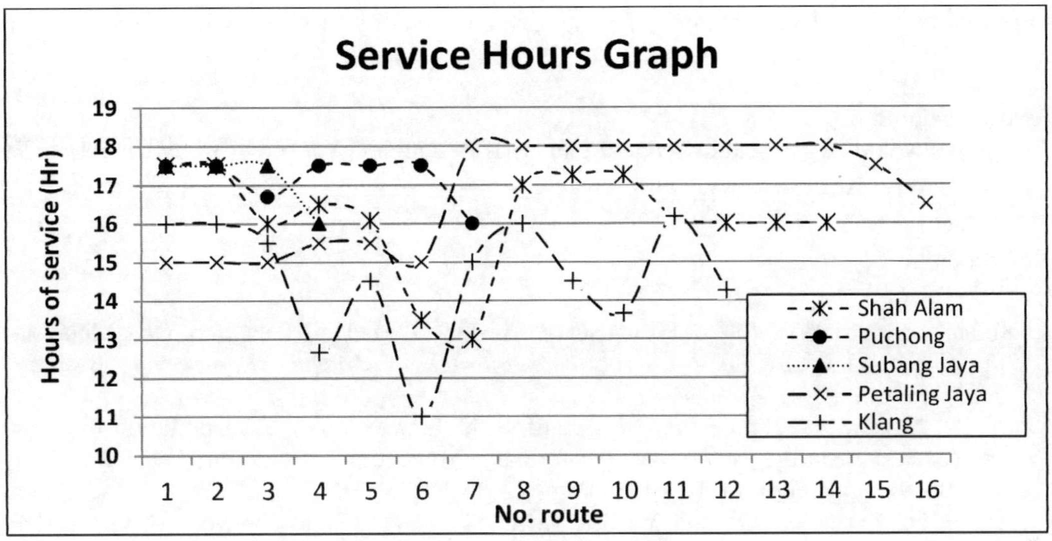

Figure 2: Service hour graphs at selected routes

For example, it can be seen that the service hours in Shah Alam were not consistent because a few routes operated at about 17.5 hours, 13.5 hours, and 16.5 hours. However, the most common operating hour for RapidKL was 17.5 hours, which started from 6.00 am until 11.30 pm. Meanwhile, for SMART Selangor,which is a free bus service provided by the Selangor State Government and operated by Prasarana Bhd, operated for 16 hours for all routes, starting from 6.00 am to 10.00 pm. In addition, the PJ City Bus, is also a free bus service initiated by Majlis Bandaraya Petaling Jaya (MBPJ) to serve only Petaling Jaya areas. It operated for 15 hours in all routes, starting from 6.00 am to 9.00 pm. Lastly, as for Seranas bus, the operating hours were not fixed just like the RapidKL buses since its service hours cover differently for Klang, which were 15.5 hours, 14.5 hours, and 15 hours, depending on the traffic flow condition, number of buses serving a particular route for that day, and the number of people who will ride the bus at that particular time. Based on Table 2, it can be seen that the QOS obtained fell into QOS B for Puchong and Subang Jaya, while QOS C was for Petaling Jaya, Shah Alam, and Klang.

4. Conclusion

The study has evaluated the service frequency for public bus service performance and service hours in Klang Valley. The results showed that Puchong had the shortest average time interval for service quality, in which the three routes served by RapidKL fell into QOS D, and Shah Alam had the longest average time interval for service quality, which fell into QOS E, whereby those routes were served by RapidKL for routes T757 (A) and T758 (B). Meanwhile Route SA01(C) was operated by SMART Selangor bus. Overall, Petaling Jaya had the highest number of routes and Subang Jaya had the lowest number of routes, which were 16 and 4, respectively. Based on the service hour analysis, the QOS obtained fell into QOS B for Puchong and Subang Jaya, while QOS C was for Petaling Jaya, Shah Alam, and Klang. The bus route has been selected by RapidKl and these routes have been approved by government's authority. Based on the results the inconsistencies of routes for each area was shown in this study and the reason for the inconsistencies is that these areas have certain difference factors such as location, living density, and transportation options. There might some room for improvement of the service quality in these areas. For example, there are some certain routes which have not been considered as main routes and competition with e-hailing services recently. In addition, several routes should be revised in order to improve the coverage area, service frequency, and service hours. The output of this study could be a reference for improving the quality service of buses in Malaysia in the future.

5. Acknowledgment

The authors would like to acknowledge the Internal Research Grant provided by Universiti Tenaga Nasional (Project No: J510050878) for this research.

6. Reference

[1] Abdullah, A. A., (2013). Rapidkl Bus Service in City Center , Kuala Lumpur , Malaysia : An Epitome of Good Service ? *International Journal of Academic Research in Business and Social Sciences*, *3*(4), 333–349.

[2] Amiril, A., Nawawi, A. H., Takim, R., & Latif, S. N. F. A., (2014). Transportation Infrastructure Project Sustainability Factors and Per-formance. Procedia - Social and Behavioral Sciences, 153, 90–98. https://doi.org/10.1016/j.sbspro.2014.10.044.

[3] Chuen, C. O., Karim, R. M., and Yusoff, S., (2014). Mode Choice between Private and Public Transport in Klang Valley, Malaysia. The Scintific World Jounal, 2014:14.

[4] Eboli, L., Forciniti, C., & Mazzulla, G. (2014). Service Coverage Factors Affecting Bus Transit System Availability. *Procedia - Social and Behavioral Sciences*, *111*, 984–993. http://doi.org/10.1016/j.sbspro.2014.01.133.

[5] Hazril, M. I., Sabri, S. M. I. & Tajedi, N. A. A. (2013). A study on previous and ongoing research regarding public transportation in Malaysia. Proc. Malaysian Technical Universities Conference on Engineering & Technology (MUCET), Kuantan, Pahang, Malaysia, 3-4 December.

[6] Larsen, O. I., & Sunde, Ø. (2008). Waiting time and the role and value of information in scheduled transport. *Research in Transportation Economics*, *23*(1), 41–52. https://doi.org/10.1016/j.retrec.2008.10.005.

[7] Leong, L. V., Ibrahim, W. H. W & Sadullah, A. F. M. (2006). Passenger car equivalents and saturation flow rates for through vehicles at signalized intersections in Malaysia. *ARRB Conference – Research into Practice 2006*, Canberra, Australia.

[8] Ma, Z., Ferreira, L., & Mesbah, M. (2013). A Framework for the Development of Bus Service Reliability Measures, (October), 1–15.

[9] Purba, A., (2015). A study on evaluating urban bus service performance in developing countries: case study of medium-sized cities in Indonesia. Ph.D. dissertation. Yokohama National University.

[10] Redman, L., Friman, M., Gärling, T., & Hartig, T. (2013). Quality attributes of public transport that attract car users: A research review. Transport Policy, 25, 119–127. http://doi.org/10.1016/j.tranpol.2012.11.005

[11] Rohani, M. M., Wijeyesekera, D. C., & Karim, A. T. A. (2013). Bus operation, quality service and the role of bus provider and driver. *Procedia Engineering*, *53*, 167–178. https://doi.org/10.1016/j.proeng.2013.02.022.

[12] Sabra, W., and A. (2003). A Comparison between Highway Capacity Manual and Critical Lane Volume Analysis Methodologies for Traffic Impact Study Applications, 1–7.

[13] Shariff, N. M., (2012). Private Vehicle Ownership and Transportation Planning in Malaysia. In *International Conference On Traffic And Transportation Engineering* (Vol. 26, pp. 64–68).

[14] Suwardo, Napiah, M. B., & Ibrahim B. Kamaruddin, (2009). On-Time Performance and Service Regularity of Stage Buses in Mixed Traffic. World Academy of Science, Engineering and Technology, 3(7), 1137–1144.

[15] Suwardo, W., Napiah, M., & Kamaruddin, I. (2008). Punctuality and Expected Waiting Time of Stage Buses in Mixed Traffic. *Jurnal Transportasi, 8*(3), 213–226.

[16] Titheridge, H. (2017). ScienceDirect ScienceDirect The attitudes of tourists towards a bus service : implications for policy from a Maltese case study. *Transportation Research Procedia, 25*, 4110–4129. https://doi.org/10.1016/j.trpro.2017.05.342.

[18] Transportation Research Board. (2013). TCRP Report 165: Transit capacity and quality of service manual (3rd ed), Transit Cooperative Research Program, The Federal Transit Administration, Washington D. C.

[19] Vien, L. L., Wan, A., & Wan, H., (2006). Passenger Car Equivalents And Saturation Flow Rates For Through Vehicles At Signalized Intersections In Malaysia, 1–9.

[20] Yaakub, N., & Napiah, M. (2011). Public Transport: Punctuality In-dex for Bus Operation, World Academy of Science. *Engineering and Technology, 60*(12), 12–23.

Light Pollution and Health: Case Study of the Lighting Fixtures Applied on Penang Town Hall in George Town, Penang Island

[1]Alaa Abdalla Saeid Ali, [2]Safial Aqbar Zakaria, [3]Lim Jit Wei, [4]Ng Yee Shien

Email: lolitta.usm@gmail.com

ssafial@usm.my

Abstract. Lighting design plays a central role in architecture and interior design. When it comes to heritage buildings, the case becomes even more sensitive and requires a delicate consideration of the light application. It is essential to conduct a good source of light to maintain a pleasant sustainable environment and to prompt a positive, healthy behavioural response. The objectives of this study are to outline the issues occur by the lighting application to the heritage building to learn the required factors to be considered in the lighting system as well as recognizing the association between the light pollution and health. The observations done around George Town on Penang Town Hall led to conduct the obstacles of the quality of light and building conservation. It is quite apparent that designers must meet the balance between the beauty of light and the appreciation of historical buildings in George Town

Keywords: Health & Light, historical building, Lighting Design, Conservation, Lighting Fixture, Art & History, Light Pollution, awareness.

1. Introduction

The importance and uniqueness value in the historical buildings in George Town - a UNESCO World Heritage Site - have been giving it the top priority for conservation, which has been applied within all aspects including lighting fixtures. However, designers, nowadays, tend to use extra light fixtures to emphasize the beauty of the heritage buildings and make it more standing for the tourists. Therefore, improper application can react negatively leading to harm those heritage buildings. This reason puts the responsibility on designers to maintain a balance in lighting design due to the extreme and unnecessary lighting application, which is leading to a negative impact, not only towards heritage buildings, but also it causes damage to humans, animals, and plants. The current paper is an attempt to present a case study investigation of the lighting fixtures applied to the Penang town hall in George Town. Thus, the methodology adopted in this research is the quantitative method theory by case study through collecting data from many related sources and analyzing it to get the final results, which are supported by personal observations, and visual data to examine the problems and questions in this research. The objectives of this study are to investigate the issues caused by the lighting applied to the heritage buildings as well as identifying the relationship between light pollution and the health aspect and finally, to study the necessary factors to consider in the lighting system. Research Method

Content from this work may be used under the terms of the Creative Commons Attribution 3.0 licence. Any further distribution of this work must maintain attribution to the author(s) and the title of the work, journal citation and DOI.
Published under licence by IOP Publishing Ltd

The researchers collected primary data through the observation method, and they allocated the site observation at the area around the City Hall in George Town. George Town is the capital city of the Malaysian state of Penang Island. It has attained UNESCO's Heritage City status and exposed to artificial lighting consideration by Penang State Government improvement plans [1]. The Penang state government has also studied it to measure the nature and extent of the building's lighting issue [1]. Observation method was applied to study the problem of lighting design at this site. The observation was issued on daytime and nighttime to define the difference between the situation of the building being let out under daylighting and artificial light. Researchers studied the characteristic of the lighting fixture of Penang Town Hall building, the lighting fixtures arrangement and the lighting system functionality along with its impact on the building and the surrounding. Then, secondary data was acquired by library research method. Secondary data collection was by collecting information from a variety of sources such as books, journal, and Internet data based. Through detecting and locating related information, and analyzing what has found, the researchers developed and expressed ideas on how to solve the lighting issues. As stated previously, the quantitative method was used to collect the data. Quantitative data was collected and each one had its importance in firming the study and assisting the situations, which were further supported by visual proof. Researchers have focused on the investigation case study of the lighting fixtures applied on Penang Town Hall. The results of the discussion and analysis will be analyzed globally, and this will be beneficial for future in-depth studies on light and health in the context of Malaysian national health. The result quantified throughout the aid of measuring instrument, which were Lux meter, luminous flux meter, and luminance meter. In summary, the researchers summarized the research methods in the following diagram.

3. Light and human well-being

Light is crucial to human health. Light radiation affects the visual cortex, and also the performance and well-being [2]. According to (Flynn, Spencer et al. 1973, Flynn and Spencer 1977, Gibson 1979), both of the physiological and psychological aspects of an individual can be boosted by applying a proper lighting design [3]. Besides that, as its been known light enables people to see and observe their environment, as well as it induces Non-Image-Forming (NIF) effects, which consequently triggers the health [4]. NIF effects vary from hormone production and cell detachment to changes in behavior [5]. Studies on inferior-nasal retinal areas proved that when using melatonin (sleep hormone) as a biomarker, it is further capable in prompt of the NIF effects of radiation [6]. Disruption of the circadian rhythms with the 24h day/night cycle can play a role in mood disorders [7], cardiovascular disease [1], and even cancer [8, 9]. As stated, radiation has long-term effects, but at the same time it, short-term effects as well, such as emission of melatonin hormone and pupillary reflex [6]. According to Boivin (1996) and Czeisler CA (1989), one of the causes for a desynchronized pacemaker effect is bright light during the night hours, which can lead to harmful health consequences, such as severe impacts on the cardiovascular system and the kidneys [10]. The harmful impacts affect the biological system, mainly when the quality of light used at night does not meet the human health necessities [11]. The researchers conclude that light is an extremely crucial element to human well-being; therefore, it is required to apply the light following accurate design principal in order to avoid the negative impact on health.

4. The relationship between light pollution and health aspect

Nowadays, light pollution has become more noticeable and seen as a significant threat by many institutions and organizations. Its occurrence is connected with adverse effects, such as sky glow, light trespass, glare, and the hindering of astronomical observations [7]. Different kinds of technical reports [12, 13], standards, and legal regulations, related to the control and reduction of this phenomenon, have recently been developed in all regions of the world. Light pollution identified as luminous pollution, is the excessive and misallocated use of artificial lighting. Light pollution is not a new term. Light trespass is one of the light pollution types and it occurs when unwanted light passes in one's property; for example, in this research light trespass could be identified as the unwanted light towards

Penang Town Hall, which reflects from the surrounding buildings. Another type of light pollution is the sky glow, which is the glow effect over populated areas. Sky glow happens when the light is produced directly into the atmosphere, accidentally or intentionally, where it is distributed by dust and gas, creating a dome-like orange glow that covers the night sky. Light pollution brings serious health issue to living beings. The circadian clock is defined as the physiologic processes effect in almost all beings. When the circadian clock disrupts, it leads to numerous medical disorders in humans, including depression, insomnia, cardiovascular disease, and cancer [14]. Serotonin is a hormone produced during the day; serotonin converts to melatonin only in darkness, while Melatonin, is a type of hormone produced during the night and its identified to be helpful in regulating the body's biological clock [15]. The body produces melatonin at night, while melatonin levels can drop impulsively in the existence of artificial or natural light [4]. Several scientific research studies indicated that decreased melatonin production leads to higher risks of breast cancer in women. On the other hand, animals can be confused by artificial lighting, leading to disrupting their breeding cycles. Moreover, the photosynthetic cycles of plants can be negatively affected when exposed to artificial light during the nighttime lights [16]. Finally, when it comes to the environment, excessive lighting causes wastage and harm to the environment. The International Dark-Sky Association states that additional lighting during nighttime wastes more than $1.5 billion in energy and electricity costs yearly around the world besides discharging of upwards of 12 million tons of carbon dioxide into the atmosphere [17].

5. Description of Heritage Building

This research is focusing on one of the most significant heritage buildings in George Town, Penang Island, which is the Penang Town Hall. British Army engineers designed the building, and it is completed in the 1880s. It is the city's oldest municipal building, as it once housed the Municipal Commission of George Town [18]. It also functioned as a venue for social events for the European elites, but in 1903, its administrative function was taken over by the City Hall, and it has been a historical monument since 1993 [6].

6. Findings and discussion

6.1. Characteristic of the visual environment

The appearance of a building depends on the visual qualities, which rely on the features of form, light, color, and materials that affect them [9]. One of the beautiful colonial buildings to appreciate is the old Penang Town Hall building. This famous two story building in George Town is known for its classical Victorian style. An impressive building situated next to the coastline with lush public gardens at the front with the value of the attractive facade and exterior that worth a look-see. A drop of shadows helps to render forms and textures of the Penang Town Hall building. The researchers' site observation leads to an understanding point about the uniform ambiance of daylight in the day and uniformly illumination during the night. In the daytime, the building is clearly letting out with the natural lighting. On the other hand, during the night time, the installation of lighting fixtures does create a focus on specific areas of the buildings. However, the arrangement of the lighting project has caused glare from the reflection on the white surface of the columns, which affects the details of the historical buildings and makes them hard to be distinguished. *Key design factors and criteria of lighting systems*
The type of lighting fixture that has been installed on the Town Hall is floodlight. The color of the projected light is white color light. According to the guidelines given by UNESCO on heritage buildings, floodlights used on the building must be natural white light. The perceived color indicates the wavelength of light being reflected. In this case, white light contains all the wavelengths of the visible spectrum therefore, when the color white is being reflected, that means all wavelengths are being reflected, and none of them is absorbed, making white the most reflective color. The full spectrum that forms white light is listed in the table below.

Color	Wavelength	Color	Wavelength
Violet	380-450 nm	Yellow	570-590 nm
Blue	450-495 nm	Orange	590-620 nm
Green	495-570 nm	Red	620-750 nm

The observation concluded that it is unnecessary to use highlight intensity floodlight to give impact on the details of the building facades only. Uniformly projected lights, such as daylight, are more than enough to enhance the beauty of the details on the facades by the contrary effect of light and shadows. The light illuminance level for the day can meet up to 9,000 lux. However, it might be slightly different in the situation at night. According to illuminance level, which is collected using the lux meter, the illuminance from the spotlight on the field shows an illuminance level of 28 lux depending on the distance from the source. The ambiance of the building was softly illuminated and faded gradient along the distance. From the researchers' point of perspective, the illuminance level of 50 lux on the area is already sufficient for the eyes to distinguish the building and the details and shadows of facades. It is undoubtable that the daylight has its uniqueness to enhance the beauty of the building beside its aspects in the virtual comfort, and then energy saving. Therefore, there are some design considerations when implementing floodlight into designs, which are the type of light source, reflectivity, color, texture of the projection surface, a distance of the surface from the light source, and color filter used. One of the sources that also affects the floodlight projection is the distance of the surface from the light. Closer distances of the surface from the light source will lead to an increase of the illuminance level towards the surface. On the other hand, the researchers suggest using Polaroid to reduce 50% of the light intensity towards the surface of the building. In order to reduce the glaring effect in addition to the eye comfort leading to differentiate the detail of the building facades, making it enjoyable without the glaring discomfort.

7. Conclusion

From the discussions above, the researchers conclude with the question of what are the characteristics necessary to successfully implement "good" light art projects for historical build necessary characteristics in George Town and, which part needs to be illuminated and which architectural element should be left dark? As it is described previously about the Town Hall, the only illuminated part during the night time is the columns, while the top of the building is dark. Therefore, the answer is somewhat complicated. In general, the layout of lights alone is not sufficient to make a building aesthetically pleasing. An analysis of the surrounding area needs to be carried out, and the illumination of historic buildings needs to be planned comprehensively, by taking into consideration how the building interacts with its surroundings and moreover expanding the concept of lighting to cover this broader scope. Thus, visual comfort and glare are two priority areas that must be addressed. It is vital to consider the environmental and biological factors that hugely have impact on the lighting effects. In the future, designers will not face the danger of ignorantly using the lighting design as it is seen nowadays as a source of inspiration, without regard to the critical aspects presented above.

References

[1] Ariffin N F M, Ahmad Y and Alias A 2015 Stakeholders' attitude on the willingness-to-pay value for the conservation of the George Town, Penang World heritage site *Journal of Surveying, Construction and Property* **6** 1-10

[2] Zakaria S A, Rou N Y, En H Z and Iyian T 2017 Colour Light And Wellbeing: A Case Study Of M Mall 020 George Town, Penang Island. In: *IOP Conference Series: Materials Science and Engineering*: IOP Publishing) p 012005
[3] Flynn J E and Spencer T J 1977 The effects of light source color on user impression and satisfaction *Journal of the Illuminating Engineering Society* **6** 167-79
[4] Boyce P R 2010 The impact of light in buildings on human health *Indoor and Built environment* **19** 8-20
[5] Boyce P R 2014 *Human factors in lighting*: Crc Press)
[6] Hockton K and Tan H 2012 *Penang: An Inside Guide to Its Historic Homes, Buildings, Monuments and Parks*: MPH Group Publishing)
[7] Żagan W and Skarżyński K 2017 Analysis of light pollution from floodlighting: Is there a different approach to floodlighting? *Light and Engineering* **25** 75-82
[8] Khademagha P, Aries M, Rosemann A and Van Loenen E 2016 Why directionality is an important light factor for human health to consider in lighting design? *International Journal of Sustainable Lighting* **18** 3-8
[9] Stevens R G 2009 Light-at-night, circadian disruption and breast cancer: assessment of existing evidence *International journal of epidemiology* **38** 963-70
[10] Boivin D B, Duffy J F, Kronauer R E and Czeisler C A 1996 Dose-response relationships for resetting of human circadian clock by light *Nature* **379** 540
[11] Zakaria S A and Bahauddin A 2015 Light Art for Historical Buildings: A Case Study of the Heritage Buildings in George Town, Penang Island *Procedia-Social and Behavioral Sciences* **184** 345-50
[12] Cha J, Lee J, Lee W, Jung J, Lee K, Han J and Gu J 2014 Policy and status of light pollution management in Korea *Lighting Research & Technology* **46** 78-88
[13] Hölker F, Moss T, Griefahn B, Kloas W, Voigt C C, Henckel D, Hänel A, Kappeler P M, Völker S and Schwope A 2010 The dark side of light: a transdisciplinary research agenda for light pollution policy *Ecology and Society* **15**
[14] Chakraborty R, Ostrin L A, Nickla D L, Iuvone P M, Pardue M T and Stone R A 2018 Circadian rhythms, refractive development, and myopia *Ophthalmic and Physiological Optics* **38** 217-45
[15] Oshika T, Klyce S D, Applegate R A and Howland H C 1999 Changes in corneal wavefront aberrations with aging *Investigative Ophthalmology & Visual Science* **40** 1351-5
[16] An C 2018 An assessment of the sharing economy and its policy solutions through the lens of sustainability
[17] Johnston J 2015 Money Talks, Sustainability Walks *Money*
[18] Kvaskoff M and Weinstein P 2010 Are some melanomas caused by artificial light? *Medical hypotheses* **75** 305-11

Architecture graduate work readiness: The gap between learning and employability.

Mari TS[1], Srirangam S[2], Gunasagaran S[3], Kuppusamy S[4] and Ang FL[5]
[1,2, 3 & 5] School of Architecture, Building & Design, Faculty of Innovation and Technology Taylors's University Lakeside, Selangor, Malaysia
[4]School of Built Environment, University of Reading Malaysia, Iskandar Puteri, Johor, Malaysia.
E-mail: TamilSalvi.Mari@taylors.edu.my

Abstract. The study investigates and compares the experience and expectation of employers and graduates on architecture graduate work readiness in relation to employability skills using a quantitative method. A total of 141 graduates and 85 employers completed a set of survey measuring the same variables but from different perspectives. The collected data was analyzed using SPSS v 25. Study findings show that perception of employers and graduates are similar on the architectural course and how it trains the graduates for practice. However there is a small gap in the perception graduates work readiness; architectural course curriculum, and graduates' practical skills. The results revealed that the graduates confidence level in their work readiness is moderate (M = 3.42., SD = 0.843), concurring with the employers who reported that graduate are not entirely ready (M = 3.62., SD = 0.648). However, the employers' score employability skills is lower than the graduates' score, especially in areas such as knowledge of Uniform Building By-Law (UBBL) requirements and sustainable approaches (RK) (M=3.29, SD=0.881) and CTS (M= 3.41, SD=0.837). Thus, the study suggests opportunities in current learning in architectural course to be transformed using "work based learning" to enhance the graduates employability skills.

1. Introduction

Employability is an important outcome of Higher Education Institutions (HEI). It is inevitable that HEI system address the employers' needs in designing the curriculum. Many literatures have reported that recent graduates lack of employability skills required by the industries. Additionally, the economic down turn has intensified the pressure on HEIs to produce graduates that has the required skills to join the work force [4]. Paper qualification alone does not guarantee employment in the competitive job market as the graduates lacks the skills required by the industry. Fadil Suhaili, Esa Ahmad Padil, and Mohamed Jamal Abidah Ainah [6] reported that lack of suitable soft skills by graduates affected their job application (p. 123). Meanwhile, severe criticism has emerged against the traditional and relatively outdated university curricula, for which the major burden of rising graduate unemployment (Fernandez-Chung & Leong, 2018). As a result, the non-compatibility of university curricula has created a mismatch between the graduate skills and the skills required by the practice or industry creating the gap between learning and graduate readiness for practice [2]. Several literatures have shown that the architectural education places the least importance on development of employability skills [22 and 21]. Architecture graduates lack the employability skills and HEIs have taken numerous efforts to embed teaching or assessment of the skills in the programme, for instance to enhance the skills development, work related

Content from this work may be used under the terms of the Creative Commons Attribution 3.0 licence. Any further distribution of this work must maintain attribution to the author(s) and the title of the work, journal citation and DOI.
Published under licence by IOP Publishing Ltd

activities could be integrated in teaching and learning [11]. There are evidences to support that work experience leads to positive outcomes [9]. The study identifies two gaps in the existing literatures: i) that there is a need for critical evaluation of the learning in architecture education and the immediate need for interventions to prepare its graduates for the real world, and ii) there are abundance of employability skills related studies mostly in other fields with very limited in the architecture studies of Malaysian context. Thus, the study seeks to investigate the readiness of architecture graduates for the practice from the graduates and employers perspectives. The study aims to compare the experiences and expectations of employers and graduates, of their employability skills and to what extent these skills are acquired through their learning in the selected HEI.

2. Current state of architecture graduates' employability skills

As a professional discipline, architecture requires its students to acquire appropriate skills and knowledge during their tenure of studies to facilitate architectural practice. Despite, the commitment by architecture schools to provide architectural education that will enable its graduates with skills related to architectural practice, graduates still lack employability skills. Similarly, report by [18] pointed out that the architectural profession has failed to capitalize on its essential capability by not creating the range of skills required to meet the demands of the present construction industry (p. 9). The Skills Survey by [17] reported that employers and recent graduates are not impressed with the skills and abilities of newly qualified architects. The study revealed few interesting alliances in the perspective of employers and students, for instance both the employers (80%) and graduates (73%) agreed that recent graduates are not equipped with practical skills required to practice architecture; more than 80% of employers and graduates agreed that graduates lack the knowledge to construct what they designed; 81% of employers and 74% of students agreed that architectural schools put theoretical knowledge above practical ability, and more than 75% of both the employers and graduates believe that architecture students should spend more time learning in practice during their studies. A Malaysian study by [19] on architecture graduates soft skills from employers' perspective, revealed communication and presentation skills in English (mean scores of 4.51 and 4.64) as most important followed by teamwork skills (mean score 3.7). Whilst the study among the architecture undergraduates by [15] revealed that students overate the level of skills and identified three (3) skills that lacked in the graduates they are; capacity for synthesis and analysis, creativity and capacity to generate new ideas, and organization, planning and time management skills. Study by [21] highlighted that sound Computer Aided Design skills and teamwork skills are highly valued discipline skills by employers at time of hiring followed by self-management (mean score 4.2), communication (4.15), creative initiative and enterprise (4.13), planning and organization (4.08), problem solving (4.0), technical (3.9), and lifelong learning (3.8). Savage, Davis, and Miller [20] pointed out that professionals emphasized on graduates critical thinking skills in addressing issues such as sustainability in their research. Falk [7] stated that employers pay high importance to graduates' readiness in teamwork (p. 4). A Malaysian study by [22] to find how the employers view the set of skills being facilitated in the design studio modules, concludes with the various discrepancies in terms of academic training and industry expectation, recommends for the universities to set their direction in studio teaching with reference to industry expectation in mind.

2.1. *The Architecture education*

The teaching and learning of the architectural education traditionally includes various learning experience to prepare graduates to real world. Over years this teaching and learning has faded due to numbers of students, logistics, facilities and others issues. Currently most of the class activities with "simulated or virtual" problems and sites do not provide students with real life experience and sense of connection to context-people and place. Author [13] criticized that current architectural education is built upon analogic reality where students are given a "virtual problem" and are requested to assume that it is a real problem (p. 86). Students generally lack the opportunity to reflect the acquired theory or content knowledge in solving practice-based issues. Authors [7] and [13] corroborate that real life projects can contribute towards development of collaborative techniques, communication skills and participatory practice and all other approaches that are vital and relevant for future architects (p. 19). WBL is an integral part of linked learning and help foster the goal of providing students with the skills they need to succeed in college and career. Thus, WBL paves solution to narrow the gap between learning and employability.

2.2. Work Based Learning (WBL)
WBL is a new educational programme, the significance of this form of education in the modern and rapidly changing society is shown in a recent examination of practice and literature indicating a growing sophistication in the way that work-based learning is being theorized and facilitated in higher education. With its gradual growth as a distinct field of practice and study with the support of relevant pedagogies and concepts of curriculum [14]. The main characteristics and essential aspects of WBL programs profitably different from traditional university programs. They are relevant to actual production processes, student centricity, flexibility of content, which is built proceeding from interests of an employer and a student. It is highly considered as a point of difference in developing graduate employability by enhancing skill outcomes amongst team-work, communication, self-management, and problem-solving, employment prospects and student understanding of the world-of-work [12]. Thus, by incorporating WBL the architecture course will have a strong foundation that will shape architecture undergraduates into architects. As criticized by [8], "relying on past and even present architecture education models is not the best strategy" (p. 70). Jackson [12] investigates the role WBL in improving undergraduate employability skills, reported a significant improvement in undergraduates' perceived ability to perform employability skills following placement. WBL is a new education programme that is increasing participation in higher education because there is support to enhance employability and upgrading skills for a competitive market [14].

3 Research Methodology
The study employed a quantitative method using survey questionnaire [5). The survey questionnaire was customized based on the HEI's graduate capabilities and employability skills. Two (2) sets of survey questionnaire was developed containing same questions but addressing two (2) groups of respondents; employers and graduates. The graduate respondents comes from a particular private HEI in Klang Valley while the employers are from architecture firms that recruits graduates from this HEI based on the list provided the HEI's Placement Centre. The graduates completed their undergraduate degree in the year 2016 and 2017 and have experience working in architectural firms. Survey questions was piloted and validated by two experts. From September 2018 to January 2019, a total of 141 graduates and 85 employers completed two sets survey measuring the same variables but from different perspectives. The collected data was analyzed using Statistical Package for Social Studies v. 25 (SPSS). Descriptive analysis (frequency, mean, scores and standard deviations) were used to analyze and explain the reported data.

4 Results and Discussion
To discuss the findings from the study, mean scoring is divided into 3 main levels of interpretation: scoring of 1.00 to 2.33 as low, 2.34 to 3.66 as medium, and 3.67 to 5.00 as high. For the purpose of discussion and analysis the scale of strongly agree (SA) and agree (A) will be summed up, likewise for the disagree (D) and strongly disagree (SD) scales. Table 1 shows the comparison of total means for architectural course and graduates readiness. Results suggest that the Graduate Readiness for employer and students are at moderate level with students scoring of $M = 3.42$ ($SD = 0.843$), while employers scored slightly higher ($M = 3.62.$, $SD = 0.648$). Thus, both opinioned that graduate readiness is not high. Efforts need to be taken by the HEI to address the issue associated with graduate readiness.

Table -1. Perception of Employers and Graduates of Architectural Course and Graduate Readiness

	Employer (N=85)		Graduates (N=141)	
	Mean	Std. Deviation	Mean	Std. Deviation
Architecture Course	3.67	0.442	3.43	0.388
Graduate Readiness	3.62	0.843	3.42	0.648

Table-2 exhibits the findings on architectural course (AC) offered by the selected HEI. Surprisingly almost 75% of both employers and graduates agreed (A)/strongly agreed (SA) that the architecture course offered by selected HEI emphasizes theoretical knowledge above practical skills. This finding concurs with the findings reported by The Skills Survey Report from RIBA Appointments (2014). The architectural curriculum assumes an adequate theory and, in turn, inevitably reveals the quality of the theory on which it is based. Ayiran ([1] argued that theory is required because it is not possible to design in an intellectual vacuum. While theory is needed, it contradicts with the essential intention of design if the theory is taken as a prescriptive principle for successful design activity. There seem to be a considerable difference among employers (77.6%) and graduates (53.2%) agreement on the alignment of the architecture curriculum to reflect current needs. The finding is not surprising, as the employers seem to have an opinion that the students may have not been exposed to adequate practical training that is critical for employment, while the graduates expect whatever they learned in school to be literally applicable in practice. In support of the above opinion almost more than 80% of employers and graduates agreed (A/SA) that the learning in AC should include real-life projects and the findings corresponds with the findings by RIBA [17]. To address this concern, majority of employers (92%) and graduates (70.2%) agree that the schools should offer internship for a minimum of six (6) months as oppose to the current duration of internship eight (8) weeks. The result can be supported by statement by [16]. Currently, the programme offer a maximum of 8 weeks of internship in the undergraduate programme. The duration is considered as inadequate by both the employers and graduates, as longer and thoughtful internship programs can help smooth that transition by better integrating the realities of client service, budgets, and building codes and etc. More than half of the employers (56%) and graduates (66%) agreed that the AC trains students to work independently, whilst almost 75% of both the employers and graduates agreed that the architectural education trains students to be more adaptable to change or the need to learn.

Table-2: Employer v. Graduates Perception of Learning and Training in the Architecture Course

Architecture Course (AC)	Employer N=85 (%)					Graduate N=141 (%)				
	SD	D	N	A	SA	SD	D	N	A	SA
Emphasizes the theoretical knowledge above practical skills	3.5	8.2	25.9	45.9	16.5	3.5	2.1	19.1	68.1	7.1
Reflects current architecture needs	0.0	5.9	16.5	49.4	28.2	5.0	12.8	29.1	46.1	7.1
Trains stds to work independently	0.0	15.3	29.4	29.4	25.9	2.8	7.8	22.7	45.4	21.3
Trains stds. to be adaptable to the need to learn	0.0	10.6	15.3	37.6	36.5	0.0	5.0	20.6	53.9	20.6
Should include more real-life projects	0.0	0.0	11.8	42.4	45.9	0.0	0.0	17.7	35.5	46.1
Should offer internship for a min. of 6 months	0.0	0.0	8.2	31.8	60.0	0.7	6.4	22.7	32.6	37.6

Table- 3 depicts employers and graduates perception of graduate readiness for practice. Contradicting to the employers (55%) only 38% of the graduates agreed that they are work ready upon graduation. It is interesting to note that this result contradicts with findings by [17] which is rather low (20-25%).

Similarly, less than half of the graduates (49%) agreed that they have adequate practical skills to practice as compared to the employers (52%). Surprisingly, opinions of employers and graduates are almost similar (more than 50%) agreed that graduates have adequate theoretical knowledge to design, construction knowledge to build what they have designed, and soft skills needed to practice architecture. These findings refute (higher) to previous results reported by [17]. These results indicate that the graduates do possess the skills but not sufficient to practice. This could be because the students were exposed to a very limited set of real life projects, issues and clients hence the opportunity for them to experiment their theoretical and technical knowledge and soft skills are limited. In general, the graduates believed they lack practical skills (49%) and they are not entirely work ready (38%) as compared to the employers who believed otherwise.

Table-3: Employer v. Graduates Perception of Architecture Graduate Readiness for Practice

Graduates Readiness	Employers N=85 %					Graduates N=141 %				
	SD	D	N	A	SA	SD	D	N	A	SA
Are "work ready" upon graduation	0.0	20.0	24.7	34.1	21.2	5.0	15.6	41.1	34	4.3
Have adequate construction knowledge to build what they design	0.0	28.2	16.5	25.9	29.4	3.5	17.7	24.1	50.4	4.3
Have adequate practical skills	0.0	20.0	29.4	25.9	24.7	1.4	14.9	34.8	44.7	4.3
Have adequate theoretical knowledge	0.0	10.6	29.4	38.8	21.2	1.4	6.4	33.3	52.5	6.4
Have adequate soft skills	0.0	15.3	21.2	38.8	24.7	0.0	5.00	31.2	54.6	9.2

Table-4 presents the mean score for employability skills demonstrated by graduates during their employment. As anticipated, the graduates tend to overestimate their skills concurring to the study by [15]. The result show that both employers and graduates mean scores is moderate for critical thinking skills (CTS) and knowledge of Uniform Building By-Law (UBBL) requirement and sustainable approaches (RK), career related and teamwork skills (CRT). Interestingly, the highest mean score for both the graduates and employers is for communication skills (COM) (M = 3.69, SD = 0.783) and (M = 3.78, SD = 0.533) respectively, thus these results contradicts the findings by [19]. Considerable differences in mean scores are observed for CRT (M=3.64, SD=0.732), CTS (M=3.41, SD=0.837), RK (M= 3.29, SD=0.888) by employers and graduates CRT (M=3.76, SD=0.557), CTS (3.72, 0.551) and RK (M= 3.60, SD=0.594). The employers scored lowest for RK (M= 3.29, SD=0.888), this due to the reason that the graduates knowledge UBBL, requirements and strategies on sustainable architectural approaches are not demonstrated during their employment. These finding perhaps can be associated with the undergraduate AC offered by the HEI is design orientated, emphasizing more on theories, construction and technology. The submission requirements, compliance such as uniform building by law (UBBL), and sustainable approaches are taught in the undergraduate degree but lacks the emphasis in the design studio modules, thus the students focus in creating designs that meet the brief and most often failing to produce holistic designs. The graduates scored lowest (M=3.50, SD=0.595) for their technical skills (TECH). Among the items they scored low are, their understanding of materials and construction (M=3.35) and skills in using Building Information Modeling software (BIM) (M=3.49). The employers also report that the graduates demonstration of CTS are at moderate level (M=3.41, SD=0.837), but the graduates believe otherwise by scoring high (M=3.72, SD=0.551). These findings corresponds with findings in [15]. Despite of having concerted effort by HEI to focus on CTS as

IOP Conf. Series: Materials Science and Engineering **636** (2019) 012010 doi:10.1088/1757-899X/636/1/012010

measurable outcome; employers are not seeing the results. Employers highlighted that CTS gap is a significant problem with new hires, specifically in recent graduates.

Table-4.0: Employer v. Graduates: Total Mean Score of Graduate Employability skills

Employability skills		Employers (N=85)		Graduates (N=141)	
		Mean (M)	Std. Deviation (SD)	Mean (M)	Std. Deviation (SD)
CRT	Career related and team work	3.64	0.732	3.76	0.557
CTS	Critical thinking skills	3.41	0.837	3.72	0.551
COM	Communication	3.69	0.783	3.78	0.533
TECH	Technical skills	3.52	0.793	3.50	0.595
RK	By law, requirement and sustainable app.	3.29	0.881	3.60	0.594

4 Conclusion

Findings from the study has revealed employers and graduates have similar perception of the architecture course and how it trains graduates for practice. Majority of both the groups agreed that the course offered by the architecture curriculum of selected HEI emphasizes on theoretical knowledge than practical skills. However, both employers and graduates suggest that more real life projects to address this concern and duration of the internship to be increased to a minimum of 6 months. The employers believe real-life projects in the curricula will equip students with essential practical skills. There is a slight gap in the perception of employers and graduates on graduates work readiness. More than half of the employers believe that the graduates are work ready compared to the graduates. Similar finding is also observed, in evaluating the graduates practical skills which is directly associated with their work readiness. Interestingly opinions of employers and graduates are almost similar on the adequacy of theoretical, construction knowledge, and soft skills. In general, the findings on employability skills of the shows that employers mean scores are generally lower than the graduates. The study highlights that employers acknowledge the school's contribution in providing relevant curriculum but not the learning approaches. In addition to, expecting graduates to have technical and discipline skills they also require graduates to demonstrate a range of skills that include communication, critical thinking, problem solving and practice related skills. Whilst, the graduates on the other hand believe that the HEI will provide the necessary training to equip them with the indispensable employability skills to ensure they are work ready. It can be concluded that both employers and graduates agree that graduates are not entirely work ready, lack practical skills, and AC emphasizes theoretical knowledge by suggesting to include more real life projects to develop their practical skills and suggest that the school to extend its internship from 2 months to a of minimum of 6 months. There is as a serious concern among the employers relating to the moderate development of the graduates employability skills during their undergraduate studies. Therefore, it is the responsibility of the stakeholders, especially HEI, employers and students to ensure the development of work related skills during their studies to ensure graduate readiness. Based on the findings and literatures, the study would like to highlight that employer and graduates value WBL (such as placements and internships) particularly as an effective approach to promote the employability of graduates. HEIs and employers should continue to promote and expand opportunities for students to access WBL and incorporating soft skills in their curricula particularly in assessment. These experiential opportunities require careful planning and time for reflection must be built in as an effective way of providing university students with relevant employment skills, knowledge and awareness of employment

Acknowledgements.
The authors would like to acknowledge Taylor's University's support of this study in providing funding through the Emerging Research Funding Scheme.

References

[1] Ayiran N 2011 Designerly way of understanding the role of theory *ARCHTHEO* **11** pp 290-300
[2] Aliyu A M 2016 Strategies for Improving Practical Projects in Woodwork in Colleges of Education (Technical) in North-Western States of Nigeria (Doctoral dissertation)
[3] Clarke M 2018 Rethinking graduate employability: The role of capital, individual attributes and context. *Studies in Higher Education* **43** (11) pp 1923-1937
[4] Crayford J, Fearon C, McLaughlin H, and van Vuuren W 2012 Affirming entrepreneurial education: Learning, employability and personal development *Industrial and Commercial Training*. **44** (4) pp 187-193
[5] Creswell J 2014. *Research design*. 4th ed. SAGE Publications.
[6] Fadil Suhaili, Esa Ahmad,and Mohamed Jamal Abidah Ainah 2015 Soft skills construct for architecture graduate in accordance with industries requirement *International Journal of Humanities, Arts and Social Sciences* **1** (3). doi: 10.20469/ijhss.20002-3
[7] Falk A 2012 Enhancing the team experience in Service Learning courses. *Journal for Civic Commitment* **18** pp 1–16
[8] Feldman R M and Grondzik W 2009 Affecting change in architecture education / can, we do the right thing. *ARCC Journal* **6** (1)
[9] Freudenberg B M Brimble and Cameron, C 2011 WIL and generic skill development: The development of Business students' generic skills through Work-Integrated Learning, *Asia- Pacific Journal of Cooperative Education,* **12 (** 2) pp 79-93
[10] Hafizah Husain, Siti Salasiah Mokri, Aini Hussain, Salina Abdul Samad and Rosadah Abd Majid. 2012 The level of critical and analytical thinking skills among electrical and electronics engineering students, UKM. *Asian Social Science*, **8** (16) pp 80–87
[11] Harvey L 2005 Embedding and integrating employability. *New Directions for Institutional Research 2005* (128) pp 13-28. doi: 10.1002/ir.160
[12] Jackson D 2013 The contribution of work-integrated learning to undergraduate employability skill outcomes *Asia Pacific Journal of Cooperative Education* **14** (2) pp 99-115
[13] Jann M 2010 Revamping architectural education: ethics, social service, and innovation. *International Journal of Arts and Sciences* **3** (8) pp 45–89
[14] Lester S and Costley C 2010 Work-based learning at higher education level: value, practice and critique. *Studies in Higher Education* **35** (5) pp 561-575. doi: 10.1080/03075070903216635
[15] Maina J J and Daful C K 2017 Do they measure up? Architecture Graduates' Perception of Acquired Skills, Employers' Expectations and what is obtained *Journal of Research in National Development* **5** (1) pp 153-161
[16] Omar N, Abdul Manaf A, Helma Mohd R, Che Kassim A.and Abd. Aziz K 2012 Graduates'employability skills based on current job demand through electronic advertisement. *Asian Social Science* **8** (9) doi: 10.5539/ass.v8n9p103
[17] RIBA Appointments 2014 Skills Survey Report (Survey) (pp. 1–28). UK: Royal Institute of British Architects (RIBA). Retrieved from ribaappoin tments.com
[18] Royal Institute of British Architects (RIBA) 2005 RIBA constructive change: A strategic industry study into the future of the architects' profession. (pp. 1–19). London, UK.
[19] Salleh R, Yusoff M A M, Harun H and Memon M A 2015 Gauging industry's perspectives on soft skills of graduate architects: Importance vs Satisfaction. *Global Business & Management Research* **7** (2)
[20] Savage S M, Davis R M, and Miller E 2009 Exploring graduate transition from university to the

workplace: employer, academic and graduate perspectives. in *34th AUBEA Annual Conference: Managing change - challenges in education and construction for the 21st century*, Barossa Valley, South Australia.

[21] Shannon S 2012 “I wish for more than I ever get": Employers’ perspectives on employability attributes of architecture graduates *Creative Education*, **03** (06) pp 1016-1023. doi: 10.4236/ce.2012.326153

[22] Srirangam S and See K.L 2017 Skills on Demand. An Investigation on industrial expectations onto fresh graduates. Taylor’s Teaching and learning Conference 2017. Kuala Lumpur: Taylor’s University

Layering of History in Malaysian Modern Architecture through Analytical Diagramming

Siti Balkish Roslan[1] and Amir Ariff [2]

[1]School of Architecture, Build and Design, Taylors University, 1, Jalan Taylors, 47500 Subang Jaya, Selangor, Malaysia

[2]School of Women's and Children's Health, University of New South Wales, 2052, Australia

E-mail: sbroslan@gmail.com

The paper looks at the Malaysian Modern Architecture buildings of historical significance from the post-independence (post-Merdeka) period of 1950s by using analytical diagrams to examine their historical layers. The 1950s beheld an important transition of architecture history in Malaysia as the newly independent nation sought to formulate an identity through various enterprises, including architecture. The usage of these historical buildings has changed over time. Considering a building as a singular entity and ignoring its past influences diminishes its historical significance and neglects history as a continuous evolutionary pressure upon the building. As architectural history develops over time, it will have layer upon layer of its physical, functional, and contextual aspects affected by factors pertaining to socio-culture, politics, and economy. Visual representation is important to capture changes and to be open in anticipation of new layers that encapsulate the building's reading at a certain point of time. The idea is to not only capture the momentous spirit of a building, but also multiple monumental and ambiguous changes that keep true to history. Such a visual representation allows an easy reading of the historical value and an open discussion through visual clues and analytical diagramming. In this study, we have selected the Modern building, *Angkasapuri*, as an archetype for interrogation via analytical diagrams.

Keywords: Malaysian Modern Architecture, analytical diagramming, layering, interpretive design learning

1. Introduction

In order to frame the contemporary debate of the usage of a modern building, a transdisciplinary approach via analytical diagramming through layering with a historical perspective is taken to explore post-Merdeka buildings that have endured notable or ambiguous transitions from one use to another or even losing relevance. The success and failure of Modern architecture and its possible

Content from this work may be used under the terms of the Creative Commons Attribution 3.0 licence. Any further distribution of this work must maintain attribution to the author(s) and the title of the work, journal citation and DOI.
Published under licence by IOP Publishing Ltd

role in the future needs to be assessed against a complex background (Frampton, 1980). Most significantly, ambiguous transitions are primarily overlooked or discounted when it might affect the building in long term. The critical examination of historiography with the spinal narratives and sporadic narratives aims to underline elements of continuity throughout the period of post-Merdeka. Continuity in this research encapsulates the process through which the building has changed, ranging from adaptation, reuse, artistic interventions, and routine daily practices that may slowly alter collective understanding of site and relevant criteria. Coupled with the said criteria, the information is abstracted into a diagram. The information becomes organizational, looking to include formal, spatial understanding and programme. The variables in diagramming later include 1) space and event, 2) force and resistance, 3) density, 4) distribution, and 5) direction. Within this architectural context, these variables transcend conventional dichotomies. The configurations that are developed are subjected to continual modification and change, showing potential relationships to be develop among elements (Allen, 1998). Were these changes ignored and the changes imposed upon the building overstep the original version; we have already lost a fragment of the historical identity of it.

There is a displacement of understanding the chronological development of nationhood in architecture in reference to the history of these modern buildings origins and how it developed up to present times within the Malaysian Modern Architectural context. In retrospect, Modern Architecture's canonical viewing centers primarily on the 'Master Architect' narrative, major movements, and exemplary buildings in Europe and North America (S.W. Lim & Chang, 2012). With this, there should be a critical perspective on the difference of early development of Malaysian Modern Architecture in terms of Eurocentric and Non-Eurocentric historiography and what it has become today, addressing both monumental and ambiguous changes. Malaysian Modern Architecture comes from a diverse and heterogeneous background and the study and re-theorization of it have made for richer debates and contents in historical subjects. The reading works of architecture from a purely historical perspective by photography and literature are not sufficient to formulate a thorough understanding of their chronological qualities and developments, specifically when pertaining to the more ambiguous changes. An analysis through historical layering of diagrams can lead to a methodical framework for its rich and subtle context to be extracted, discussed, interpreted, compared, and ultimately read for their contemporary use in interpretive design learning. This opens a way of new readings of the buildings that seeks to contribute to interpretation (Radford, Markoc, & Srivastava, 2014) and, iterating the importance of analysis in understanding methods of learning architecture historically from a design perspective.

We cannot discount the origins of Modern Architecture and its European past. However, this foundation is forced into the historical understanding of Malaysian Modern Architecture, when earlier it was argued on its diversity and heterogeneity. Malaysian architectural education is exposed to literature of canonical western works (Curtis, 2005), but not only is there a gap in relating the formal and spatial qualities analyzed in canonical Western works to the works of Malaysian Modern Architecture from the same period and framework but also in understanding the chronological historical transitions of the buildings themselves. In this paper, we look at the main government telecommunications building, Angkasapuri, as a reference for its significant chronological changes and how it was influenced and affected by contextual factors.

This leads to a gap in the research where influences over the building's lifespan within the context and content of the building are discounted, as there are in the assimilation and interpretation or re-interpretation in the works that were built during the post-Merdeka period. It relies on factual recollection concerning historical and significant individuals through interviews and recordings involved in the conception of the building. Layering analytical diagramming aims to capture a chronological reading of historical value of the building through visual clues, understanding architecture history from a design point of view, and encouraging interpretive design learning application.

2. Methodology

2.1 Identification and Literary Investigation

A Malaysian Modern building was assigned to the student group. The students identified the building as is for its formal and spatial architectural understanding without accounting for the influences of the context. This is to identify criteria of Modernism, measure it against the Western tenets, and compare it to Malaysian Modernist tenets. The students then find literature to analyze the history of the building, in this case, *Angkasapuri*.

The literacy and photo evidence of post-Merdeka buildings are abundant, yet the dissection of information in terms of design language are few. Prior to identification of buildings that have gone through changes in terms of adaptation, reuse, artistic interventions, and routine daily practices that may slowly alter collective understanding of site, understanding of the evolution of Eurocentric Modern Architecture is explored. This is to formulate the connection between Western-centric Modern Architecture and its arrival and transcendence into Malaysian Modern Architecture and its role of development in Malaysia's nationhood. The background literature forms the basis of investigation; data collected is used for an in-depth knowledge of the subject of study. The advantage of the method allows for a synthetical comprehension as well as availability and accessibility of information as having an accurate and thorough representation of findings to analyze (Roslan & Anuar, 2018).

2.2 Field Studies

The students proceeded to visit the site of *Angkasapuri* to collect information unattainable via literary means. Informal interviews were conducted with employees and exploration of the building was done. First-hand visits were essential for acquired knowledge. From here the development of sketch maps, measurements, and drawings, recordings, and field notes that frame conjectures, in such a narrative, achieves the weight of an informed opinion (Groat & Wang, 2013).

The process encourages further comprehension of interpretive analytical thinking towards the contextual architecture that succinctly looks at the adaptation, reuse, artistic interventions, and events that alter the collective understanding of the site and building chronologically. In addition, the main basis of information such as the architect, context, climate and structure to arrive to the conclusion of a layered diagram.

2.3 Layering and Diagramming

Diagramming at base understanding is a form of representation and condensed information in graphical form to show in depth relation of the design. It gives a different perspective, observation, and analysis, which at times are not represented well when other methods and mediums are used. Architectural diagrams apply a varied range of graphical indicators; for example; topology, shape, size, position, direction, emphasis, imagery, and others and these are subsets of elements and principles (Clark & Pause, 2012). In the case of this exercise, we look at a combined method using layering of graphical indicators comprising of elements and principles.

The base of the diagram, as shown in figure 1, is used as a point of reference and drawn on a muted opaque sheet of paper. The following diagrams, as shown in figure 2 – 5, in relation to the highlighted criteria are drawn on a separate sheet of paper that has a translucent quality. When all diagrammed criteria are combined and placed on top of one another, this shows a layered analytical diagram of condensed information of the history of the building.

The ideal outcome would be a singular diagram of a building that houses multiple layers of representation of historical changes. From here, the layering of history in analytical diagramming allows a visual representation supporting a proposed analysis, focused idea, and an abstraction for detail (Do & Gross, 2001). However, it can be argued that this method has an imprecise quality, but allows the exercise to have an interpretive design approach to a usually rigid methodology of learning history.

2.4 Interpretive Design Learning

Interpretive design learning and research was used. It intended to develop a form or method of documenting in analyzing and representing historical changes of the building; the students started with the collected data of *Angkasapuri* and documented the chosen Modern building through analytical diagramming. The students explored the building and identified ambiguous and hidden information behind a complex, interrelated, or multifaceted context and for the construction of an analysis in areas with no or insufficient priori information. The traits of the following are also taken into consideration; history bringing in a view from the past, interpretation, narrative, and considering a cultural turn (Groat & Wang, 2013).

As the research looks at the collected data to formulate a layering of Modern Malaysian Architecture history through formative and spatial ideas and also adaptation, reuse, artistic interventions, and routine daily practices that may slowly alter collective understanding of site, it is appropriate for researching context-specific, unique, and idiosyncratic subjects, events, and processes. This helps uncover compelling and thought provoking research questions and issues for continuous research and follow up.

IOP Conf. Series: Materials Science and Engineering **636** (2019) 012011 doi:10.1088/1757-899X/636/1/012011

3. Results and Discussion

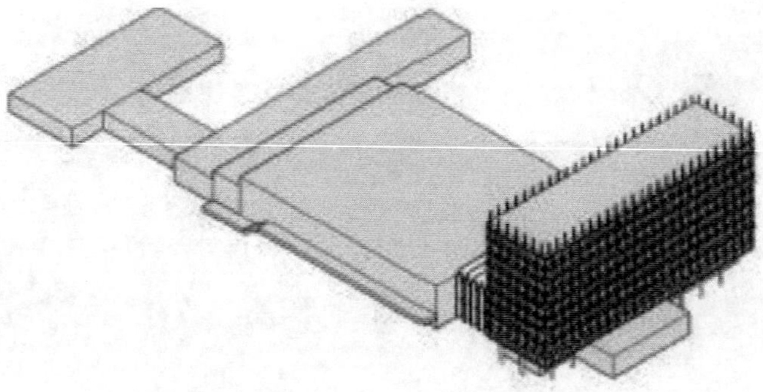

Figure 1 *Base of the axonometric diagram of Angkasapuri developed by Cheong Yen Sin*

The findings show, historically, *Angkasapuri* is multi-layered, elaborated, and complex in comparison to a singular image of its physical composition and aesthetics. The findings carry more information when combined with the contextual understanding chronologically throughout the lifetime of the building and reactions to socio-culture, political, and economic influences.

In interpreting collected evidences of the past of the building, the technical aspect of interpretation is applied to project a view and narrative through the prescribed methodology. As a result, it illustrates a layered analytical diagram that possesses hermeneutical qualities that allow for future conceptions of discussion and learning.

The example given in figure 1 shows the axonometric base of *Angkasapuri* in its original form from 1968, highlighting key principles, elements, and design of the building to be assessed and analyzed. In figures 2 and 3, the layering through analytical diagramming shows the documentation of a design understanding that states the growing and expanding of *Angkasapuri* through chronological expansion. Though completed in 1968, the broadcasting operations moved its office to the *Angkasapuri* Complex, which began its first telecasting in 1969. It is also important to note that this expansion is also due to the merging of radio and television under the Ministry of Information at the time. It is clear that the influences that lead to continued building and extension were due to the climatic change in socio-culture, politics, and economics in relation to the establishment of a growing nation. This is a result of higher demands and varying activities pertaining to distribution of information. *Wisma* Radio, Engineer Department, and Auditorium were built in the second stage in 1972, continued with the International Broadcasting Center completed in 1988 and finally *Wisma Berita RTM* completed in 2012.

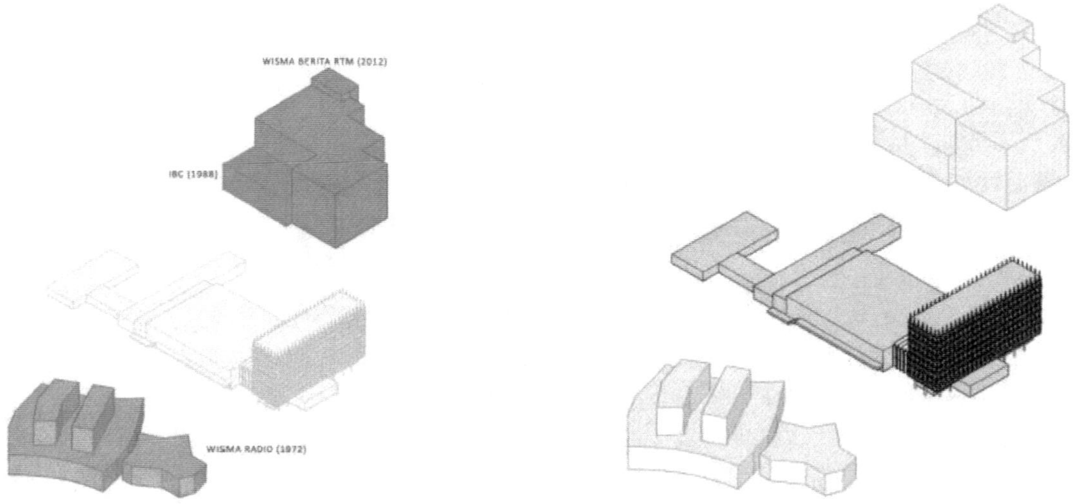

Figure 2 and 3 *Layering of history through analytical diagramming (chronological extension and adaptation) developed by Cheong Yen Sin*

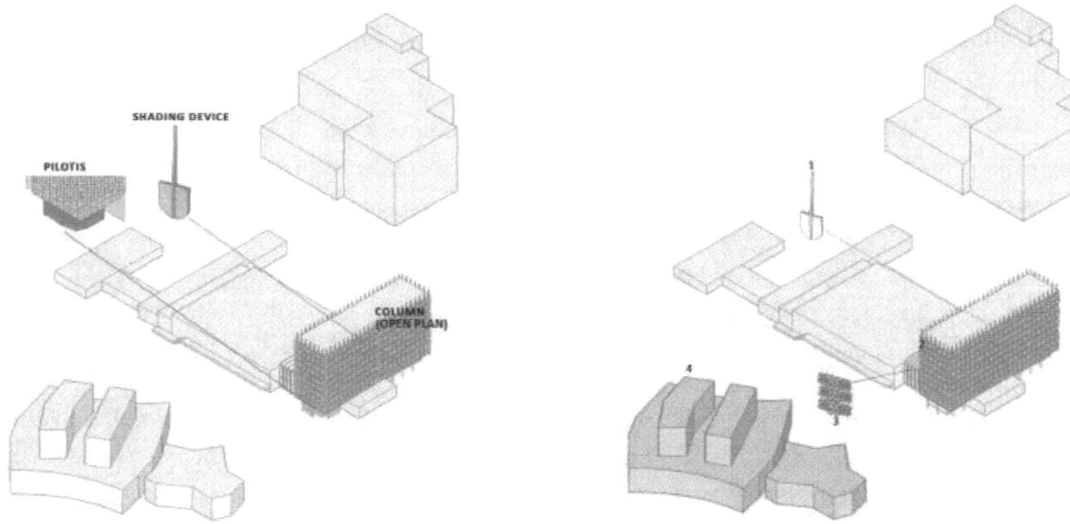

Figure 3 and 4 *Layering of history through analytical diagramming (artistic intervention and space layout and usage) developed by Cheong Yen Sin*

4. Conclusion

The layering in history of Malaysian Modern Architecture was an attempt at a methodology that encourages interpretive design learning. The value of design lessons from Malaysian Modern Architecture projects had little implementation in practice in the subsequent years due to lack of documentation of the processes and design intentions underlying their conception (Tajuddin, 1999). It allows the subject of the history of the building to be described and acknowledged to have an overlooked existing reality through multiple changes over a lifespan, whether monumental, ambiguous, or having less relevance. There is also a space of discussion of the building for a probable reality that has yet to exist through tracking the layering of diagramming history chronologically over the years in identifying a repeating or predictive pattern. For Modernism outside of the Western realm, there will inherently be an intertwining of nationalism, regionalism, globalism, colonialism, and other influences that define the understanding and definition of the context of the country involved. Through the methodology of interpretive design learning, there is room for the re-defining of Malaysian Modern Architecture through the relevance of geopolitical, economic, and socio-cultural aspects without having to revert to the Western tenets of Modern Architecture, taking into account the convolution of influential aspects (regionalism, nationalism, globalism, colonialism) and modernism.

Historically, architecture history classes in Malaysian universities have always adopted Western tenets as a basis for teaching. Through observation, this might be the case of globalization, and lack of local knowledge and materials to formulate a syllabus or being unable to justify the importance of the transaction of local history within the Malaysian architecture.

What many fail to understand is that modernism is not a homogenizing process leading to a diversified socio-cultural fabric adapting to different contexts that converge towards a banal monolithic Modernism in reference to the tenets of the West. Rather, the processes of Modernism have led to a multiplicity of social and cultural formations that have confronted, contested, and diverged from the formulations of the West's understanding and coining of what Modernism is in a Malaysian context. The prescribed tenets can be questioned and its implications of divergence do not mean that Malaysian Modern Architecture is any less authentic and regarded as any less Modern, hencelayering of history of Modern Malaysian Architecture through analytical diagramming is crucial to allow an in depth understanding of layered Modernism.

5. References

[1] Allen, S. (1998). Diagrams Matter. van Berkel and Bos.
[2] Clark, R. H., & Pause, M. (2012). *Precedents in Architecture-Analytic Diagrams Formative-Ideas and Partis.*
[3] Curtis, W. J. R. (2005). *Modern Architecture Since 1900* (3rd Edition). Phaidon Press, New York.
[4] Do, E. Y. L., & Gross, M. D. (2001). Thinking with diagrams in architectural design. *Artificial Intelligence Review*, *15*(1–2), 135–149. https://doi.org/10.1023/A:1006661524497
[5] Frampton, K. (1980). *Modern Architecture: A Critical History*. Thames and Hudson.
[6] Groat, L., & Wang, D. (2013). *Architectural Research Methods* (Second). Wiley.
[7] Radford, A., Markoc, S., & Srivastava, A. (2014). *The Elements of Modern Architecture: Understanding Contemporary Buildings*. Thames and Hudson.
[8] Roslan, S. B., & Anuar, A. N. (2018). Analytical Diagramming as a Pedagogical Tool in Understanding Malaysian Modern Architecture (p. 12). Selangor: Pertubuhan Akitek Malaysia (PAM).
[9] S.W. Lim, W., & Chang, J.-H. (Eds.). (2012). *Non West Modernist Past*. Singapore: World Scientific Publishing Co. Pte. Ltd. https://doi.org/https://doi.org/10.1142/8233
[10] Tajuddin, M. (1999). *Seni Bina Di Malaysia: Kritikan Tentang Seni Bina Islam, Identiti Nasional & Pendidikan. Design Modular.*

Acknowledgements

Thanks to the Architecture and Culture History Class 2 (ARC 60203), semester March 2019.

IOP Conf. Series: Materials Science and Engineering **636** (2019) 012012 doi:10.1088/1757-899X/636/1/012012

Chinese Cemetery as Heterotopia: A Case of Kwong Tong Chinese Cemetery (KTCC)

V Ng and M Goh

School of Architecture, Building and Design, Taylor's University, 1, Jalan Taylor's, 47500 Subang Jaya, Selangor, Malaysia

Ng.foongpeng@taylors.edu.my

Abstract. With the emergence of new typologies of memorial parks and commercialisation of industries for the dead in Malaysia, what is the image and fate of older cemetery sites within the urban realm? This paper aims to re-define the perception of urban Chinese cemetery as a heterotopic space through the case study of Kwong Tong Chinese Cemetery (KTCC). Using the Foucauldian notion of heterotopia as a framework for analysis, this paper studies the transitions of KTCC through time, and how it stands in relation to the society that produced it. The paper argues that the cemetery is a highly heterotopic site; it is a space of representation and a space of difference, which celebrates the disruption of space and time in relation to society. The implication of society towards the ever changing role of a cemetery, and for now, the changing progression is at the stage where the contemporary Chinese cemetery in Malaysia displays its multi-functional spatial qualities within a single site, as a space of 'otherness'.

1. Introduction

This paper aims to re-define the perception of urban Chinese cemetery as a heterotopic space. Using Kwong Tong Chinese Cemetery (KTCC) as a case study, it discusses the ever changing function, role, and condition of urban Chinese cemetery in Malaysia through the theoretical position of heterotopia by Michel Foucault. This position arises from the contestation of old urban cemeteries, the perception of fear and taboo related to them. Recent literature has shown that the cemetery has been both exceptional and separated from everyday regularities and at the same time, a domestic, or just an ordinary space. In Malaysia, prior to the development controversy over the historic Malaysian Chinese cemeteries, it did not carry much political influence, except as burial grounds for deceased Chinese [1]. They have been largely neglected and abandoned, except during Ching Ming festival, which happens once a year where people pay their respects to the deceased. However, Loo said that the Chinese cemetery is a space that relates to the collective memory of the Chinese community about their cultural roots and their sense of belonging to Malaysia [2]. This question of roots is both cultural and political; more importantly it is also spatial. Current days, cemeteries surrounded by the encroaching and congested cities may function as urban green belts; new commercial cemeteries are seeking methods to increase the function and human presence in cemeteries by providing new form of physical spatial elements, such as garden, café, jogging track, or community hall. All these elements place the Chinese cemetery in a highly contested space through time [3]. The current context of the

Content from this work may be used under the terms of the Creative Commons Attribution 3.0 licence. Any further distribution of this work must maintain attribution to the author(s) and the title of the work, journal citation and DOI.
Published under licence by IOP Publishing Ltd

Chinese cemetery raises a fundamental question of how do contemporary conditions of cemeteries shape how we define and think about them. The perception of cemetery as a contested space links to the theory of other space or heterotopia, as the notion of heterotopia is to argue how space and place are always agents for processes of subjectivity, and stand in a political relation to the society that produced them [4]. This paper explores heteropology in a contemporary Chinese cemetery in relation to Foucault's principles of heterotopic space, and subsequently, examines the relational spatial quality in a contemporary Chinese cemetery in Malaysia.

2. Literature Review

2.1. The Chinese Cemetery and KTCC

The Chinese cemetery in Malaysia existed since the settlement of Chinese, especially during the period of British colonial rule from the late 18th century onwards. Whether they are the traditional or commercial cemeteries, their primary role is to serve as the "house" of the dead and to facilitate the practice of the burial culture. According to Loo [2] and Tan [1], the cemeteries have not been established merely to provide this utilitarian service. Not only do they bear historical witness to the role of the Chinese development of Malaysia, but also create a deep cultural and historical imprint on the landscape, especially the old Chinese cemeteries in Malaysia, such as Bukit Cina in Malacca and Kwong Tong Chinese cemetery in Kuala Lumpur. They are now a permanent component of the built environment in Malaysia and constant adaptation towards new management practices apply to each traditional cemetery as a functioning cemetery through time [5].

In Malaysia, the former Chinese cemetery is approached in the context of the core notion of "unity of heaven and humans" and the latter in terms of the management of cemeteries. Conceptually, the linkages are affected through the enactment of key cultural values of filial piety, the practice of Feng Shui, and the principle of 'Yi' [2]. Although the discourse on cemeteries has been directed to focus on environmental ethics among the dead, nature and humans, contemporary cemeteries, such as Nirvana Memorial Park and Xiao En Nilai Memorial Park in Malaysia yield insights on green culture without neglecting the traditional elements that constitute a Chinese cemetery, such as Feng Shui and the principle of 'Yi'. In practical terms, changes are being made with the aim to influence the general perception of the cemeteries as landscapes of fear and taboos, into a beautified garden of remembrance, and at the same time, to safeguard the sanctity of the burial culture and tradition [1]. Serious attempts are taken by cemetery management committees to improve the image of the conventional cemetery, to adopt alternative modes of burial, to develop commercialized memorial parks, and to nurture a green culture of cemeteries. Cemeteries in contemporary period are acquiring a new meaning that integrate an ancient practice in accordance with the more enlightened view of the environment [3].

Kwong Tong Chinese Cemetery (KTCC) is one of the oldest cemeteries located in Bukit Petaling, Kuala Lumpur, Malaysia. Founded since 1895, it covers an approximate land area of 343 acres and house to final resting place of many notable pioneers in Malaysia. There have been at least three major attempts by the state to relocate the KL Chinese cemetery, in 1994, 1998, and 2000. During 1994, the state announced the intention to develop the KL cemetery with an urban development plan [2]. The existing cemeteries were to be removed to a new designated site. After the Chinese communities raised their worries and concerns about the removal, no further plans were revealed. Following with a few more attempts from various developers, who approached the management of the Chinese cemeteries to acquire their land, the sign of contestation rises each year following the property value gain.

After the development threat, the Chinese community re-appropriated the KL Chinese Cemetery to represent Chinese history, to construct Chinese memory and at the same time, to provide an alternative interpretation of national identity and national landscape through three components – buildings, memorials, and community participation – contribute to the contestation of the state-dominant ideology, memory and history [2]. Learning from the landscaping and beautification project in the early stages in 1998, the management of Kwong Tong Cemetery combined most of the ideas, buildings, and monuments together to turn the cemeteries into a cultural and heritage park, called KL Kwong Tong Cemetery Heritage Park (KTC Heritage Park). In various ways, the cemetery is constructed in the 'Chinese' ways, and all physical and spatial elements show an expression of a form of 'Chinese-ness'. The cemetery acted as a living cultural archive for Chinese history, and at the same time, represented Malaysian history. Therefore, the Chinese-styled cemetery environment cannot be read as an enclosed Chinese world, but a symbolic spatial struggle against the state's culture. This, however, opened up a different interpretation of Malaysian nationalism, identity, and landscape: A different form and a different style, represented in stand of a political relation to the society that produced them. This could relate to heterotopia as the space and place are always agents for processes of subjectivity.

2.2. Heterotopia and the otherness

The concept of heterotopia challenges us to question the reality of what society perceived as normal and different. Michel Foucault, being the proponent of heterotopia, argued for the otherness of space, relating the principles of heterotopia against the spaces and real sites, such as the garden, brothel, cemetery, asylum, and a boat [6]. This concept remained relevant till today amongst architects, urban planners, and social anthropologist [7], due to Foucault's lack of any strict definition, leaving to the researchers a vast field for translation [8].

In explaining heterotopia, Foucault [9] spoke about the juxtaposition of hierarchic assembling of contrasting places: sacred place and profane place; protected place and open place; urban place and rural place. Foucault argued that the space, in which we live, happens to be a heterogeneous space, a space full of relations and one could describe different sites via the cluster of relations that allows them to be defined. Normal and 'other' space' co-exist: Without the 'normal' space there would not be the 'other' space, and it is the society that justifies the laws, rules, and familiarity to form a 'normal space', therefore, creating the 'other space'. According to Foucault, a heterotopia is the manner, in which society and culture, have power on the one hand and the interest of realizing this power on the other; define the subject through his differentiation from the general society [9].

From there, Foucault further elaborated two main types of sites: a 'utopia' and 'heterotopia'. He compared such ambiguous space with the tradition of 'utopia', a place (topos) that is both nowhere (au-topia) and a good place (eu-topia), in which the word 'heterotopia', came from a different or another (heteros) place: Utopia is a place in which society presents in a perfected form, but fundamentally unreal, similar to the imaginary animals and encyclopaedia; Heterotopia, a represented, contested, or inverted site that is real and effectively enacted, juxtaposed to utopia as an imaginary site presented by society itself.

3. Material and Methods

This research used case study research as a strategy of inquiry. A case study is an observation and experiencing inquiry that investigates a contemporary phenomenon or setting within its real-life context, especially when the boundaries between phenomenon and context are not clearly evident [10]. The primary data was comprised of the original literature on the theory of Heterotopia by Foucault, and followed by the adaptations, interpretation, and conception by three authors: Michiel Dehaene and Lieven De Cauter [11], and Edward William Soja [12, 13, 14]. They are selected as they all adapted different school of thoughts while translating Foucault's heterotopia yet emphasized on a real spatial quality that could have relations towards spaces in architecture. Based on review of literature, heterotopia celebrates the discontinuity and changeability of existence, and is much about the disruption of time and space in relation to the society. It could be argued that all spaces consist of certain degree of heterotopia, and Foucault's six principles of heterotopia provides a way to determine the 'level' of heterotopias in various spaces through the summarization in Table 1 below.

Table 1. Question derived from the six principles of heterotopia as a tool [15].

	Principles	Questions
1	Deviation	Is the place privileged, sacred, or forbidden, reserved for

		individuals who are, in relation to the human environment in which they live, in a case of crisis?
2	Function	Does the place have a certain function as a place within a society?
3	Diversity of spaces in a single real place	Are there series of places in the place that are foreign to one another? What these places are and how they are juxtaposing each other?
4	Time	Is the place accumulating time? How is it possible to notice the accumulation of time?
5	Place: Public-private relationship	Is the place open for everybody? What user groups is intent?
6	Society	What kind of population is operating in the place? What creates the uniqueness of the population? In which way the society in cemetery could be illusionary for random passers-by?

Each principle of heterotopia is used to determine the level of heteropology as to argue Foucault's statement that the cemetery is a highly heterotopic space. This theoretical base is used to analyse the observations of Kwong Tong Chinese Cemetery (KTCC).

4. Findings and Discussion

4.1. KTCC as a heterotopic space

Drawing from the six principles, how is the Chinese cemetery a heterotopic site? By interpreting Foucault's six principles of heterotopia into KTCC, the findings suggest that heterotopia was consistently spatialised in the context of a Chinese cemetery. All principles held a certain characteristic that were easily linked to each other in terms of time, space, and society. However, time was highly represented and involved within each principle: through 'time' the society has changed; through 'time', the space is diversified within a single place; through 'time', the function was converted; through time, the space was opened up from private to public space; and through time, the events took place within KTCC have shaped its purpose to adapt to the 'new' time. Thus, the fourth principle relating to time has the most dominant factor compared to the six principles. Although the chronological timeline was emphasized through the events, issues, and change of function, purpose and programmes in KTCC, the sequential timeline was not the focus, but the spatiotemporal relationship that occurs within the site. In the fourth principle, the accumulation of time juxtaposing with the consistently flowing and transitory time both occur within a single real site that contributes to the spatiotemporal paradox, which Foucault states in the heterotopia of a cemetery that has an ever changing function and role, which is contested, inverted, and represented as a different form through time.

The sixth principle regarding the aspect of society was repeatedly mentioned too. The case of crisis, the diversity of places, the private/public aspect, and the purpose-programme relationship were consequences relating to the society as well. For example, the society under threat of relocating in KTCC, forged spaces and programs to function within the site as an action to resolve impending issues. The examination of the situation of people existing on the margins of society is one of the main pillars of Foucault's work. He often focused on the 'negative structure' of society or an excluded group [16], which in KTCC, is the society within a Chinese community relating to a particular cemetery in Kuala Lumpur. This shows that KTCC was planned for a special group of people with a special function, making it a privileged and forbidden place as mentioned in the first principle, yet it is opened to the public as analysed in the fifth principle.

Analysing the public-private relationship of KTCC in the fifth principle, the site is blurred between pockets of private and public space. For example, the site provides private 'dead space' that was linked to a public access. This situation could relate to Dehaene and De Cauter's statement that heterotopia has the ability to be neither public nor private, occupies an 'in-between' space, which includes institutions that are fundamental to urban-life [17].

4.2. Spatialisation of KTCC as a heterotopic space

To what extent does the findings imply towards the spatial quality of the contemporary Chinese cemetery in Malaysia? Firstly, heterotopia is an experience that can be contemporary and the conception of the cemetery is related to how it stands in a political relation to the society that produced it. The analysis of the Chinese cemetery describes an experience encountered by various societies through time. Cemeteries are not just communal 'resting place' for the dead; they are also, by their very nature, ground for enactments of commemoration and places of remembrance loaded with social meanings for the living. At the same time, cemeteries are an integral part of life, reflecting society's attitudes and perceptions of both individuals and the Chinese community at large. The ability for a contemporary Chinese cemetery to be a highly heterotopic place depends on the society that governs it. The acceptance by the society decides the level of 'otherness' and the level of normality. The society, whereas in a situation of a contemporary Chinese cemetery, the Chinese community decides whether the cemetery is to be highly religious or secular, places of despair or places of hope and reconciliation.

Secondly, the juxtaposition of diverse spaces in a single space opens up a new form of cultural space. The argument here is that the cemetery in its actuality reveals a certain utility of function, but at the same time acts as a grid of intelligibility, exposing an ensemble of processes that contribute to what Foucault calls the 'art of government'. In some ways, the analysis of the emergence of the contemporary cemetery has some resemblance to the birth of the museums within a cemetery. Based on the third principle of heterotopia, the diversity of spaces in a single real space shows that a heterotopic site is able to juxtapose in a single space, several incompatible spatial elements, such as the museum in a cemetery and a cemetery in a garden. The museum does not enclose or sequestrate dangerous population or communities, however, it opens up a new form of cultural space that draws in a wider range of society and embraces a moral register and pedagogic role, and same goes with the case of a contemporary Chinese cemetery, where the garden overlays the purpose of a cemetery that was perceive its core function to be disposing the dead. The opening up of new spaces within a space, and the acceptance by the society work as heterotopian dispositives, concentrating and illustrating a whole ensemble of relations involving 'coagulation, support, reciprocal reinforcement, cohesion, and integration' (p 23) [18].

In Foucault's essay 'Of Other Spaces', he related cemeteries as a highly heterotopic space due to its ability to incorporate a number of incompatible concepts of spaces within its single framework of a physical space [19]. For the cemetery, these somewhat contradictory themes involve many aspects; for example, from the state of the body buried into the ground, the period of the body was to buried, the visiting space and events as a respect to the deceased, the funeral rituals, and the garden that encapsulate the graves within the cemetery.

Foucault explained that heterotopias are not like utopias, which are 'sites with no real space'. Utopias are unreal sites with ideas about how society should be [18]. However, the notion of utopias is not opposed to heterotopia. Heterotopias are spaces that contain utopic notions, but are different from all the other sites that they are connected to. In this example, the contemporary cemetery holds the notion of a utopic idea, realised within a real site, which could be state that the contemporary Chinese cemetery is a heterotopic place. What makes the contemporary Chinese cemetery interesting as a heterotopia, is that in addition to the layers of meaning of space that the site could interpret through the six principles of heterotopia, the notion of heterotopia in a cemetery itself could be directly translated from Foucault's essay 'of other space'.

"The heterotopia begins to function fully when people are in a kind of absolute break with their traditional time; thus, the cemetery is indeed a highly heterotopian place, seeing that the cemetery begins with that strange heterochronia that lass of life constitute for an individual, and that quasi eternity, in which he perpetually dissolves and fades away. The population, individually and collectively, becomes viewed as dependent on a 'series of variable' including material surroundings, customs, tastes, values, conduct, fears, prejudices, and means of subsistence." (p 182) [18]

5. Conclusion

This study has demonstrated the conception Chinese cemetery in Malaysia as a highly heterotopic space, a place of 'otherness', which is governed within the social conditions that shape its spatio-temporal continuum. The cemetery is a place in process. Heterotopia, through this study, is able to define the cemetery as a space of difference and a space of representation: a space in which the difference between the spatial quality of the dead and the living is placed within the same real site, made available for public contestation. Foucault's initial explanation of heterotopia in a context of a cemetery was useful, as he highlighted various issues faced by cemeteries in general and how the progressing function is constantly altered through time, which leads this study to realize the implication of society towards the ever changing role of a cemetery, and for now, the changing progression is at the stage where the contemporary Chinese cemetery in Malaysia displays its multi-functional spatial qualities within a single site, as a space of 'otherness'. The cemetery studies were diverse and relational towards the testing of the notion of heterotopia against the juxtaposing spatial elements within a single site, and seeing to what extent the framework of heterotopia through the six principles is able to 'read' the spaces and define the spatial qualities of a Chinese cemetery in Malaysia.

6. References

[1] Tan K 2011 *Spaces of the Dead: A Case from the Living* (Singapore: The imprint Ethos Books)
[2] Loo Y M 2013 Architecture and urban form in Kuala Lumpur: race and Chinese spaces in a postcolonial city (Farnham, Surrey: Ashgate)
[3] Tong X F 2003 Revisiting the Relations between Humans and Nature *Discourse on Culture Chinese Culture and Globalization* ed Fei Xiaotong et al (Nanjing: Jiangsu Educational Press) (in Chinese)
[4] Åsdam K 1995 *Heterotopia - Art, pornography and cemeteries* [cited 22 October 2016]. Available from http://knutasdam.net/images/uploads/text/heterotopia_k_asdam.pdf on 22 October 2016
[5] Rugg J 2000 Defining the place of burial: what makes a cemetery a cemetery? *Mortality* **5** (Taylor & Francis Ltd, University of York) [cited March 13, 2017]. Available from http://www.d.umn.edu/cla/faculty/jhamlin/4960/Defining%20Place.pdf
[6] Foucault M 1984 [1967] *Of Other Space: Utopias and Heterotopias. Architecture /Mouvement/ Continuité.* Translated from the French by Jay Miskowiec
[7] Johnson P 2006 Unravelling Foucault's 'different spaces' *History of Human Sciences* **19** [cited 20 October 2016]. Available from http://hhs.sagepub.com/cgi/content/abstract/19/4/750
[8] Schmidt J and Wartenberg T E 1994 *Foucault's Enlightenment: Critique, Revolution, and the Fashioning of the Self Critique and Power* ed M Kelly (London: MIT Press) pp 283-314
[9] Foucault M 1970 [1966] *The Order of Things* (Andover, Hants: Tavistock)
[10] Groat L and Wang D 2013 *Architectural Research Methods* (United States: Wiley)
[11] Dehaene M and De Cauter L 2008 *Heterotopia and the City* (London and New York: Routledge)
[12] Soja E 1989 *Postmodern Geographies: the reassertion of space in critical social theory* (London: Verso)
[13] Soja E 1995 Heterotopologies: A Remembrance of Other Spaces *Citadel-LA Postmodern Cities and Spaces* eds S Watson and K Gibson (Oxford: Blackwell) pp 13-34
[14] Soja E 1996 *Thirdspace* (Oxford: Blackwell)
[15] Einla L 2009 The Notion of Heterotopia in the Practice of Landscape Architecture *The Garden of Education – Alnarp* (The Swedish University of Agricultural Science, Faculty of Landscape Planning, Horticaulture and Agricultural Space)
[16] Potts D L 2011 *Knowledge and Power in Foucault* (Objectivism, Atlas University, Deeper Dive Archive)
[17] Dehaene M and De Cauter L 2008 The space of play: towards a general theory of heterotopia

Heterotopia and the City, eds M Dehaene and L De Cauter (London and New York: Routledge) pp 75-87

[18] Foucault M 1998 [1966] The Thought of the Outside *Aesthetics, Method, and Epistemology: Essential Works of Foucault Volume 2*, ed J D Faubion (London: Penguin) pp 147- 171

[19] Foucault M 1998 [1967] Different Spaces *Aesthetics, Method, and Epistemology: Essential Works of Foucault Volume 2*, ed J D Faubion (London: Penguin) pp 175-185

Acknowledgments

We wish to acknowledge that the study is conducted as part of the dissertation studies in the Master of Architecture programme, Taylor's University. We would like to thanks the academic staff from the programme who has contributed to the development of this study namely Professor Robert Powell, Dr. Francis Lin Chia-Hui and Dr. Sucharita Srirangam.

IOP Conf. Series: Materials Science and Engineering **636** (2019) 012013 doi:10.1088/1757-899X/636/1/012013

Integrated space in the traditional *Pondok* school education system - A case study in Pattani, Thailand.

Mohd Jaki Bin Mamat[1], Md. Baharuddin Abdul Rahman[2]

School of Housing, Building & Planning, Universiti Sains Malaysia[1]
School of Education, Universiti Sains Malaysia[2]

mohdjaki@usm.my[1], mdbahar@usm.my[2]

Abstract. The *pondok* school system was the earliest in the Malay Archipelago that focused on Islamic studies. The teaching method which stressed on group communication was said to have given rise to integrated space within a building thereby influencing its architecture and spatial design. Two of such schools, almost a century old in Pattani, Thailand, have been used as case studies to measure the influence of communication and integrated space on the interior layout. Data from interviews congruent with floor plans is the crucial reference as to the significance of integrated space. Studies have shown that emphasis on communication in the pedagogy of the traditional *pondok* school is the main creator of the integrated space, and this gave priority to the positioning and size, and subsequently influencing the overall floor plan.

Keywords; pondok institution, integrated space, communication

1. Introduction

The *pondok* school adopts a traditional methodology that accentuates communication to achieve its goal, that is, to mould the human attitude, apart from its delivery of education contents based on Islamic principles (Shuhaimi, 2013). Inevitably, this gave rise to the design of the premises' spatial layout with integrated space that functioned as a common area for various activities and involved its usage for groups within one place (Shatha, 2017).

The pedagogy method that practises communication between teacher-and-pupil and pupil-to-pupil is a primary emphasis for integrated space in the floor plans of *pondok* schools (Arslan, 2017; Yusuf, 2015). With the main building connected to other spaces in the two case studies of Pattani, Thailand, the relationship between communication and integrated space is a yardstick to its influence in the design of the interior. It also discloses the overall design of floor plans in both schools.

2. Literature Review

2.1 Pondok education institution

Generally, traditional Islamic education in the Malay Archipelago falls into two categories: the *madrassa*, which has study activities during the day, and the *pondok* which is a boarding school with full day-long studies (Nilan, 2009). 'Pondok' is derived from the Arabic *fundug* which means a hostel or hotel (Shuhaimi, 2013). The *pondok* institution is the region's system of schooling that focuses on the

Content from this work may be used under the terms of the Creative Commons Attribution 3.0 licence. Any further distribution of this work must maintain attribution to the author(s) and the title of the work, journal citation and DOI.
Published under licence by IOP Publishing Ltd

Quran and prayers based on the classical textbook (Eka, 2007). Other subjects such as science, *fiqh*, theology and Sufism are also taught (Bradley, 2013). As such, the *pondok* is considered a traditional education system complete with boarding facilities created centuries ago (Eka, 2007).

Islamic education is not just about knowledge but encompasses cultivating the attitude and human spirit (Shuhaimi, 2013). Thus, it needs to comprehend the teaching methods and provide suitable spaces based on this. Studying in the *pondok* is a full-time affair which also includes praying with the brethren. This practice is not confined just to the study place like the main hall or class, but also include other spaces like the kitchen, dormitory, prayer hall and so on (Yusuf, 2015). Therefore, Islamic education encompasses the formal which are scheduled lessons and evaluations and informal which refers to practice and lifestyle through perception, behaviour and life's lessons.

In Malaysia, Islamic lessons begin in the teacher's house and progress to the mosque and *surau* (minor mosque) when there is increased capacity. In Indonesia, the institution was called *pesantren*, noted to have emerged in the 13th Century (Eka, 2007). The *pondok* came into being from a combination of the hut (pondok) linked to the teacher or religious scholar's house and drew pupils from the villages and later from elsewhere, as well as other countries. The full-time studies necessitated pupils to build hut-accommodation around the scholar's house which became the main building in the complex (Shuhaimi, 2013).

The ministry of education in Thailand has defined the *pondok* as a 'private school with Islam instruction' (Chidchanok, 2016), and in the modern context is classified as informal education (Thithimadee, 2018). It emerged in the 14th Century during the Pattani Malay kingdom, and after the latter's fall to Siam (old Thailand) in 1786, the role of the mosque and *pondok* became increasingly important in uniting and spreading the relevant knowledge to the Pattani community (Bradley, 2013).

External factors such as economy, politics, culture and society also affected the study environment in its education (Ucan, 2018), and these determined the features of the region's *pondok* institution which have evolved since its emergence. Nevertheless, it has met the contemporary criteria which include the pedagogy method through human interaction, teacher integrity and the development of the human spirit. This is because the *pondok* school focuses on behaviour and daily practices which shape a person's lifestyle later in life (Yusuf, 2015).

2.2 Integrated space

Integrated space is defined as a shared space for specific activities, whether of the same or different nature (Chan, 2018). It can undertake a variety of such and is an open and flexible space (Arslan, 2017; Skogland, 2017). The sharing of it can bring about reciprocatively behaviour and practice, subsequently creating a social bond even though it may have been initially among strangers.

Space-sharing can create a sense of belonging (Skogland, 2017; Jarvis, 2011) and improve the exchange of knowledge, non-verbal communication and awareness (Gray, 2018). Such places are equated to social spaces where interaction take place, thus bringing about social by-products (Sara, 2017). It shows the formation of the student's social consciousness which is the aim of the studies and thus leads to the influence of the space design including the physical movement in the buildings (Montgomery, 2008).

Skogland (2017) also emphasised that the space design was crucial as it was to serve the specific activity as planned beforehand; hence, a direct relationship between space and function. Indeed, the design can affect the user in terms of interaction, movement and behaviour. In space planning, integrated space is the most important, primary and largest amongst other spaces that exist together with it (Shatha, 2017).

Its multi-purpose function also helps to reduce the number and size of other spaces designated in a building.

The study space is a joint premise for teaching and learning. The teaching area refers to the space planned for periodic meetings between teacher and pupil while the learning space is allocated for individual or group study activity (Beckers, 2016). Education is also geared towards building social skills through group study and informal interaction (Montgomery, 2008). Learning through engagement within a group justifies an individual's presence and projects his personality in the surroundings and space he is in (Pipitone, 2017). In a space, exchange of knowledge is through interaction, and so, influences its design. Knowledge-transfer is via two main channels, explicit and tacit, the latter being motions and gestures. Both require face-to-face meetings (Meulenbroek, 2017).

This is in line with Islamic education which often emphasises face-to-face meetings (Maimun, 2010), and the *pondok* adopts this traditional pedagogy where a teacher sits in front with the pupil facing him in a semi-circle. Learning this way influences the shape and size of the space by incorporating some references in the space planning of the overall building such as land location, the main entrance in relation to the road, including orientation such as facing towards Mecca for the building with a prayer area (musolla/madrassa). Given that the *musolla* is linked to the study space, the optimum would be square-shaped according to Islamic teachings whereby the brethren can pray together.

As such, the integrated space exists as a result of shared activities and at the same time being aware of the importance of communication in learning. The *pondok* school blends this area to achieve its goal during lesson time. The relationship between the importance of the integrated space and communication is gauged through location and space allocation as compared to other areas in the floor plan of the interior of the school's main building.

This research has used two *pondok* schools as its case study: Pondok Hj. Ismail bin Ahmad situated at Semela Village (in Kerisik, Pattani) built in the 1940s, and Pondok Hj. Ismail bin Hj. Idris at Bendang Badam, Yarrang District (Pattani, Thailand) built in 1947. Via interviews with the owners of the building and former pupils, information was obtained regarding the original function and features of each space. Measurements were carried out to derive physical data as to the position and size of the spaces in the main building of both schools. The data was used to show the level of significance of integrated space and compared to interview data to reflect communication activity while lessons were conducted.

Figure 1. Pondok Haji Ismail bin Ahmad, Kampung Semela, Kerisik, Pattani.

Figure 2. Pondok Hj. Ismail Bin Hj. Idris, Kampung Bendang Badam, Daerah Yarrang, Pattani, Thailand.

3. Analysis & Findings

Figure 3. Floor Plan Pondok Hj. Ismail Bin Hj. Idris

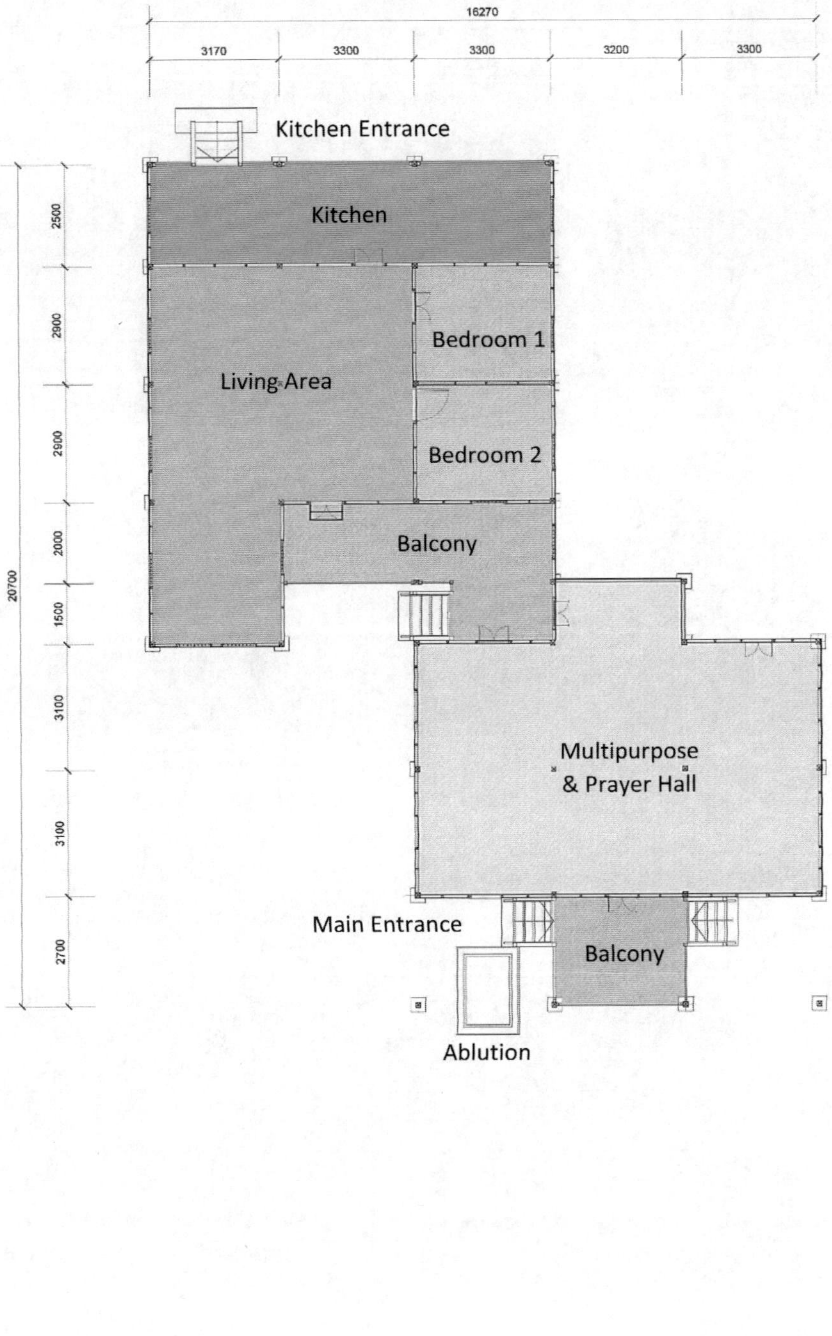

Figure 4. Floor Plan Pondok Haji Ismail bin Ahmad

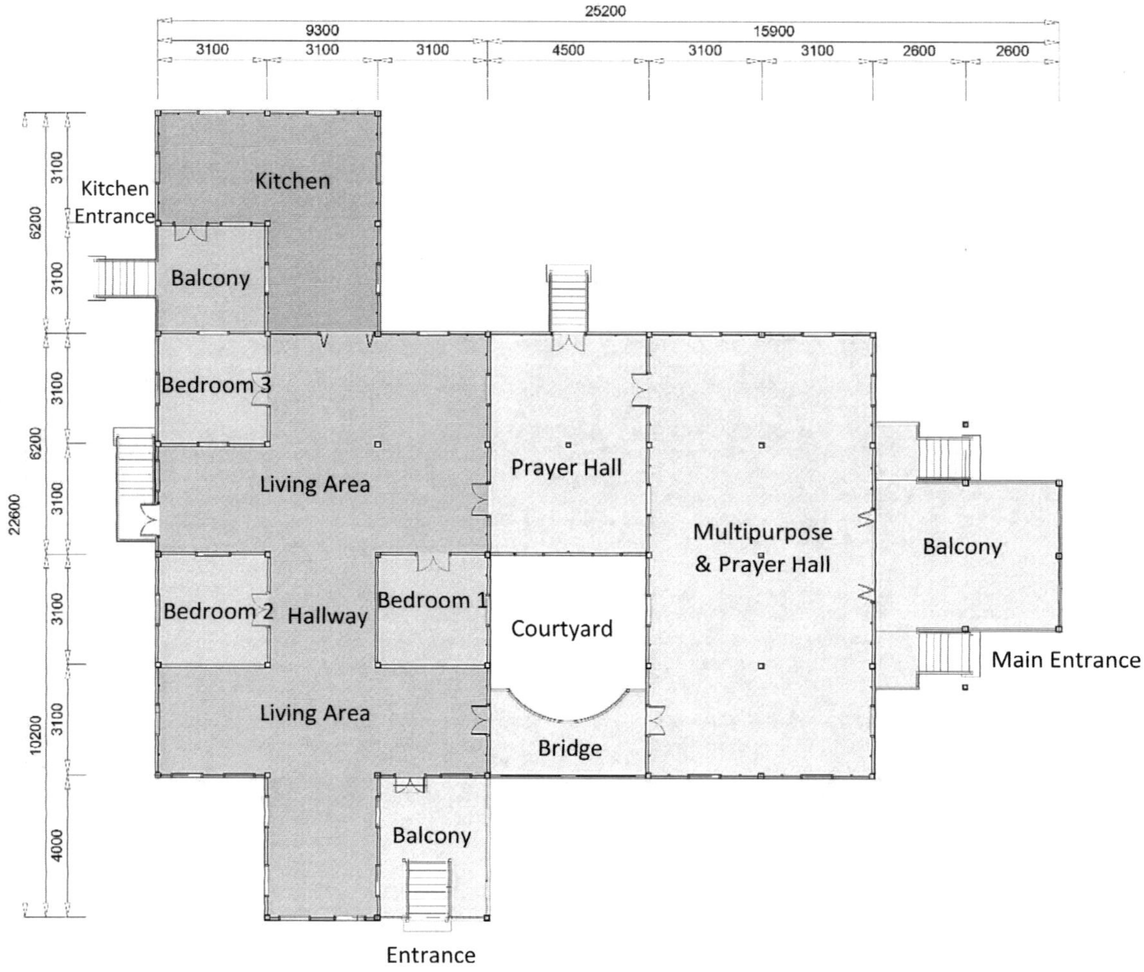

Figure 5. Size of respective spaces of the two case studies

Space	Semela Village	%	Bendang Badam Village	%
Balcony	54.24	17.2	27.79	15
Prayer & Multipurpose Hall	104.78	33.2	65.56	35.3
Living Area	89.25	28.3	48.63	26.2
Bedroom	28.8	9.1	19.14	10.3
Kitchen	28.8	9.1	24.43	13.2
Hallway	9.6	3	-	
Total	315.47	99.9	185.55	100

From the interviews it was found that the Prayer and Multipurpose halls as well as the Living Area, were integrated spaces and reflected the features and function of the schooling system in both schools; the Balcony, although an interaction area, was merely for conversation and rest. Other spaces such as the

Kitchen were considered a semi-private area and mostly used by female students, while the bedroom was for the teacher and his family - in this context the house owner.
Based on data collected, both schools gave top priority to the integrated space in Pondok Hj. Ismail bin Hj. Idris (Bendang Badam) the Prayer and Multipurpose halls together had 33.2% of total space area - the largest. Besides prayers, space was also the main venue for lectures, lessons for big groups, and for pupils to discuss and converse. It also served as a reception place for guests. At 28.3% of total space, the Living Area was second largest. Differing from the traditional Malay house where guests were usually entertained in the Living Area, in the *pondok* it was an interaction space and for small group studies. It was also there that seniors, as instructed by the teachers, usually taught the junior pupils in small groups. Combined, the Prayer, Multipurpose halls and Living Area, the integrated space of this *pondok* school has 61.5% of the overall area. The same can be seen at Pondok Hj. Ismail Ahmad at Semela where the three pertinent spaces catering to similar activities and functions encompassed an area of 61.5% also.

The integrated space is centrally located at the front. It is in the path and main circulation of the users in the building, thus clearly justifying its vital importance in floor plans. The Prayer and Multipurpose halls, for example, are situated forefront at the main entrance/door facing the road, guided by its function as an assembly area for pupils from the surrounding *pondoks* who enter the main building. It is also the area to receive visitors from villages and elsewhere that come at specific times to attend religious talks, thrusting the need for a large space at the front.
Therefore, the priority given to integrated space is evident in the two *pondok* schools mentioned which stressed on communication in pedagogy. It is in line with Islamic education which makes communication and interaction a main learning process mechanism, be it between teacher and pupil, or pupil and pupil, or with that of visitors. As such, the parallel relationship of integrated space with communication of the users of the building can be seen in the floor plans of the *pondok* school which espouses the traditional teaching and learning methodology.

4. Conclusion

It is concluded that integrated space runs in tandem with communication and determines its size and positioning in floor plans. For the two schools cited here the position at front and centre coupled with the fact that this space was the largest suggests that it is the most significance feature, hence proving that communication and interaction play a vital role in the design of the area.

Interaction and communication have been the determining factors in the floor plan design of the *pondok* school where the function of the integrated space is described in the literature review. To this, emphasis on communication and interaction in traditional Islamic education is clearly evident and practised in both schools under study here. Without a doubt, the *pondok* schools in Pattani adopt a systematic pedagogy that fulfil the education theory as well as having a well-planned interior layout that aid scholastic achievement in Islamic studies, and that is, the shaping of the student's knowledge and social consciousness.

5. References

[1] Ayça Arslan, Türkan Ulusu Uraz 2017 *Small House Spatiality: a Comparative Space Syntax application* (Open House International) vol.42 no.2.

[2] Ayse Demirel Ucan & Andrew Wright 2018 *Improving the pedagogy of Islamicreligious education through an application of critical religious education, variation theory and the learning study mode* (British Journal of Religious Education) DOI: 10.1080/01416200.2018.1484695.

[3] Chidchanok Churngchow & Withada Sinprajukpol 2016 *Factors affecting the scholastic achievement of Prince of Songkla University students from private schools with Islam instruction in the three southern border provinces* (Kasetsart Journal of Social Sciences) 37, 59-65.

[4] Eka Srimulyan 2007 *Muslim Women and Education in Indonesia: The pondok pesantren experience* (Asia Pacific Journal of Education) Vol. 27, No. 1, pp. 85–99

[5] Helen Jarvi 2011 *Saving space, sharing time: integrated infrastructures of daily life in cohousin* (Environment and Planning A) volume 43, pages 560-577.

[6] Francis R. Bradle 2013 *Sheikh Da'ud Al-Fatani's Munyat Al-Musalli And The Place Of Prayer In 19th-Century Patani Communities* (Indonesia and the Malay World) Vol. 41, No. 120, pp. 198–21.

[7] Jeffrey Kok Hui Chan and Ye Zhang 2018 *Sharing Space: Urban Sharing, Sharing a Living Space, and Shared Social Spaces* (Space and Culture) 1 –13.

[8] Jennifer M. Pipitone and Chitra Raghavan 2017 *Socio-Spatial Analysis of Study Abroad Students' Experiences in/of Place in Morocco* (Journal of Experiential Education) Vol. 40(3) 264 –278.

[9] Juliet Kerico Gray, Melissa Burel, Marlee Graser, Karen Gallacci 2018 *Applying spatial literacy to transform library space: a selected literature review* (Reference Services Review) Vol. 46 Issue: 2, pp.303-316.

[10] Maimun Aqsha Lubis, Melor Md Yunus, Mohamed Amin Embi, Sabariah Sulaiman, Zamri Mahamod 2010 *Systematic Steps in Teaching and Learning Islamic Education in the Classroom* (Procedia Social and Behavioral Sciences) 7(C), 665–670.

[11] Mari Anna Chatarina Skogland, Geir Karsten Hansen 2017 *Change your space, change your culture: exploring spatial change management strategies* (Journal of Corporate Real Estate) Vol. 19 Issue: 2, pp.95-110.

[12] Mohd. Shuhaimi bin Haji Ishak & Osman Chuah Abdullah 2013 *Islamic Education in Malaysia: A Study of History and Development* (Religious Education) Vol. 108 No. 3.

[13] Muhammad Yusuf & Ismail Suardi Wekke 2015 *Active Learning on Teaching Arabic for Special Purpose in Indonesian Pesantren* (Procedia - Social and Behavioral Sciences) 191, 137 – 141.

[14] Pam Nilan March 2009 *The 'spirit of education' in Indonesian Pesantren* (British Journal of Sociology of Education) Vol. 30, No. 2 , 219–232.

[15] Rianne Appel-Meulenbroek, Bauke de Vries and Mathieu Weggeman 2017 *Knowledge Sharing Behavior: The Role of Spatial Design in Building* (Environment and Behavior) Vol. 49(8) 874 – 903.

[16] Ronald Beckersa, Theo van der Voord and Geert Dewulf 2016 *Why do they study there? Diary research into students' learning space choices in higher education* (Higher Education Research & Development) Vol. 35, No. 1, 142–157.

[17] Sara Ibrahim Nafi' 2017 *Towards integrated socio-cultural space in "Fareej Old Al Hitmi" neighborhood in Doha* (A Master Thesis, Qatar University, ProQuest 10599867).

[18] Shatha Malhis 2017 *The Spatial Logic of Mamluk Madrassas: Readings in the Geometric and Genotypical Compositions* (Nexus Network Journal) 19:45–72.

[19] Thithimadee Arphattananon 2018 *Multicultural education in Thailand* (Intercultural educatIon) Vol. 29, no. 2, 149–162.

[20] Tim Montgomery 2008 *Space matters Experiences of managing static formal learning spaces* (Active Learning In Higher Education) 9(2).

[21] Tutin Aryanti 2013 *Breaking the wall, preserving the barrier: gender, space, and power in contemporary mosque architecture in Yogyakarta, Indonesia* (Dissertation of Doctor of Philosophy in Architecture, University of Illinois, UMI 3614537).

IOP Conf. Series: Materials Science and Engineering **636** (2019) 012014 doi:10.1088/1757-899X/636/1/012014

Adaptive reuse approaches of Shophouses at Cannon Street in George Town, Penang.

N Abdulhameed[1], M J Mamat[2] and S A Zakaria[3]

[1,2,3]School of Housing, Building and Planning, Universiti Sains Malaysia, Penang
E-mail: [1]noran.abdulhameed@student.usm.my, [2]mohdjaki@usm.my and [3]ssafial@usm.my

Abstract. George Town has many heritage buildings as a UNESCO World Heritage Site that need to be preserved due to their status. Adaptive reuse of heritage buildings is a common method of building conservation and should be done in maximum retention with minimal intervention by applying the right materials with the appropriate approach. This paper focuses on the four approaches of adaptive reuse that were applied to Shophouses at Cannon Street in George Town, Penang. The data was collected from a survey on ten Shophouses in Cannon Street, Penang. The findings of this research point to the common approach of adaptive reuse in Shophouses that has been used as the Preservation approach.

1. Introduction

Adaptive reuse of historic buildings is not a new process, however, it has been used since the French Revolution in 1790, and it was necessary to confiscate all the church's properties, buildings, lands, and works of art (Plevoets & Van Cleempoel, 2014) Adaptive reuse is a method of conservation that allows a building to be altered for a new or different use. With the world population doubling in recent decades and the enormous increase in environmental threats to our heritage buildings, our society's concern about their survival, conservation and reintegration and how to pass them on to future generations is more acute than ever (P. Bullen & Love, 2011).

The critical reflection of the use of cultural heritage buildings and sites, however, is a completely new date (P. A. Bullen & Love, 2009; Plevoets & Van Cleempoel, 2011). Adaptive Reuse provides specific opportunities for the public to perceive and attract newborn spaces to the value of historic buildings. Most experts regard adaptive reuse as the most appropriate and the only solution to preserve the legacy of the past (Pimonsathean, 2002). Heritage buildings, according to Australia's Department of Environment and Heritage, provide valuable insights into the past and characterizes communities (Reuse, 2004), hence, it should be preserved for future generations.

Conservation of heritage buildings helps to explain the past and contributes to future generations and gives a sense of continuity and belonging to the place where people live. In Malaysia, buildings designated as heritage buildings are believed to be over 50 years of age and should be preserved, protected, and upgraded in order to prevent them from being lost forever (Mustafa & Abdullah, 2013). After being listed as a World Heritage Site by the United Nations Educational, Scientific and Cultural Organization (UNESCO) in 2008, George Town is required to preserve all its heritage buildings. It is usually recommended and inspired to use adaptive reuse methods for heritage buildings to bring new life to such buildings in order to maintain the idea of the Living Heritage City. For that reason, it's prohibitive to do any physical changes to the structure of the buildings if there's any conceivable means of protecting it in its original or current condition. This is conjointly stressed in ICOMOS Burra Charter article two that explicitly states that the essence of heritage conservation in Penang is mostly retention with minimum intervention (Icomos, 1999). One of the famous Streets known for shop house in George Town is Cannon Street, which got its name from the cannons fired during the 1867 Penang Riots.

Content from this work may be used under the terms of the Creative Commons Attribution 3.0 licence. Any further distribution of this work must maintain attribution to the author(s) and the title of the work, journal citation and DOI.
Published under licence by IOP Publishing Ltd

2. Literature review

2.1. Adaptive Reuse

According to Encarta English (U.K.) Dictionary, adaptation means something modified to suit different conditions or purposes. Adaptive reuse in architecture refers to the process or state of change that fits a new environment or condition, or the resulting change. Adaptive reuse refers to suitable functional conversions of heritage buildings to suit newly proposed uses and coexist in a different environment from the original. It is the act or process of preserving something in order to keep something alive through either modernization (rehabilitation to the original function), or complete conversion to a new function or even a mixture between the two (Yildirim, 2012). Adaptive reuse is defined as a "successful marriage" between cultural heritage facilities and modern architectural design (Reuse, 2004). Also, it is described as the transformation of buildings or sites in order to adapt to modern tendencies of sustainable development (Brooker & Stone, 2008). Moreover, buildings have often changed their purposes from ancient times to fit into contemporary social needs.

2.2. Adaptive Reuse Approaches

In some cases, the adaptive reuse process may exceed the existing structure's boundaries or may even require the construction of an annexe building depending on the project's peculiarities (Eyüce & Eyüce, 2010). In other words, adaptive reuse involves any intervention to adapt, upgrade, introduce new services, and uses, while protecting the location in order to meet the desired functional requirements. The process itself should be applied to the building, while retaining its structure, character, original identity, and general authentic meaning for future generations (ElKerdany, 2002). There are several design strategies that are ambiguous whether they are projects that are reused or just new buildings that refer to historical elements. Otherwise, in adaptive reuse buildings, there are many common identifiable interventions. In addition to taking a historic structure and renovating it for future use, some approaches for reuse involve only specific elements of the original structure. For example, when the original structure would be too expensive to renovate, the incorporation of only certain building parts into a mostly new structure can be used.

A facade is maintained to retain the exterior appearance, but the entire interior structure is replaced. That means the building maintains the contextual character of the environment it is in, but it is given a completely new function on the interior (Kim, 2018). This is popular among developers as a practical solution, but not among those who seek to preserve a city's history. For a building that is too poorly repaired or a site with little historical value, a façade can be a compromise. Although there are aesthetic and cultural-historical reasons for preserving the architectural value of the building, the practice of "Facadism" is often seen as a compromise between the investor, who needs to satisfy the needs and standards in construction, and the competent protection services, which want to preserve buildings and its cultural and historical significance. "Facadism" can also be seen as a compromise between the preservation of the architectural heritage and its demolition, and therefore, is equally both praised and contested (Kim, 2018).

The *renovation* is the most common refurbishments approach in adaptive reuse projects. It is mainly used to replace deteriorated parts of the building. In many cases, the basic structure and external appearance of the original building are maintained, but the inside is completely different (Ozik, 2006). Reuse projects that fit into this category are often either pedestrian outcomes resulting from a need for more space or physical manifestations of the polemic of a designer. On the other hand, the polemical projects can be charged quite emotionally and are not usually the result of spatial or economic necessity; rather, they are the result of social, cultural, and historical commentary. As such, the interventions they propose are concise, often affecting much more than their actual fabric in reading an old space

Integration means the addition, and it is positioned next to the original building and does not interrupt the older structure in Bollack's notion, which is obviously legible; "no blurring of boundaries, no transfer of architectural elements" (Bollack, 2013). The most successful adaptive reuse projects fall under the category of integrated infill, in which much of the skin and structure of the original building is kept unchanged, while the internal space configuration is changed, sometimes significantly.

Preservation is how to reuse a historic building with the aim of preserving or restoring its original state. In addition, it is about using the existing old structure to protect original buildings, while inserting new spaces into the older volume (Bollack, 2013). According to Bollack, who describes this approach as maintaining the old building facade as a skin to preserve memories and emotion, the actual preservation defines the new space. In addition, it is emotionally attractive to the public to see and perceive such old structures and forms, as this helps preserve the existing atmosphere of the historic districts.

3. Research methodology

This research focuses on the quantitative method of using survey, which is a questionnaire that is a reliable and quick method to collect data from multiple users in an efficient and timely manner. The Personal interview was used to get data about the approaches of adaptive reuse. This survey had been done in ten-heritage Shophouses in order to identify the different approaches to the Shophouses at Cannon Street in George Town.

The survey covers a general question about the buildings as well as the approach that applied. The research sampling method used in this study was 10 users whether they were owner or renter of Shophouses In George Town. Canon Street in George Town was chosen as the location of those Shophouses due to the importance to be one of the oldest streets in George Town. Cannon Street is among the popular tourist destinations in Penang and has emerged as a commercial area since 1867.

Figure 1. Cannon Street, George Town, Penang.

Table 1. The 10 buildings in the study have changed their intended use after the adaptation process.

Building No	Types of ownership	Year build	Previous function	Current function
B1	Rental	1920's (by Khoo Kongsi)	House, Townhouse, Royal house	Commercial building
B2	Rental	1920's (Owned by Khoo Kongsi)	Townhouse	Commercial building
B3	Rental	Around (the 1890s-	Shophouse	Souvenir shop

IOP Conf. Series: Materials Science and Engineering **636** (2019) 012014 doi:10.1088/1757-899X/636/1/012014

		1910s)		
B4	Rental	1920's	Townhouse	Shophouse and townhouse
B5	Rental	More than 100 years old and owned by Khoo Kongsi	Art (teaching) dance physical	Refreshment shop (sells Ice cream, etc.)
B6	Rental	Around (1890s to 1910s)	House (residential)	-Retail shop - Refreshment shop
B7	Rental	Built around 40 to 50 years ago	Unknown	Commercial building
B8	Owned	Maybe exist before the 2nd world war.	Unknown	Commercial building
B9	Owned	Happen to pass down till now- 4th generation	Unknown	Unknown
B10	Owned	4th generation	Unknown	Unknown

4. Data analysis

Four approaches have been studied, which were 'Preservation, Renovation, Facades and Integration'. As shown in table 2, 7 out of 10 Shophouses had used the preservation approach in a significant rate by reusing a historic building to preserve or restore its original condition. While 9 out of 10 Shophouses have used the renovation approach with low rate due to replacement of much of the building that could be in poor repair. In many cases, the original building's basic structure and external appearance are maintained, but the inside is entirely different. Lastly, 3 out of 10 of Shop Houses used integration approach with the lowest rate, and it was by building around the original structure in such a way that it remain almost the same, but a new building is included.

Table2. Approaches rate on each Shophouses

Building No	**Methods /Types of Adaptive Reuse**			
	Preservation	Renovation	Facades	Integration
B1	90%	10%	0%	0%
B2	0%	10%	0%	90%
B3	20%	80%	0%	0%
B4	50%	10%	0%	50%
B5	80%	20%	0%	0%
B6	90%	10%	0%	0%
B7	0%	20%	0%	80%
B8	0%	100%	0%	0%
B9	80%	20%	0%	0%
B10	100%	0%	0%	0%
Total	51%	28%	0%	22%

5. Main findings

The main findings show that the most common approach of adaptive reuse applied is the preservation as appears in figure 2. Overall, about 51% used the preservation approach; on the other hand, 28% used the Renovation approach, whereas only 22% used Integration approach.

However, these interventions should be evaluated in order not to conflict or falsify the heritage building's uniqueness. This criterion evaluates any changes, additions, and extensions made to heritage buildings as part of their reuse adaptation in terms of respecting their authentic style, character, and identity. The main indicator is to avoid the conflict between the additions and the heritage building in architectural style.

In the end, this research paper sought to simplify a different kinds of adaptive reuse approaches that are used to Shophouses in George Town. Ultimately, the data obtained can guide and assist the heritage sector as well as the designers in developing new adaptive reuse approaches.

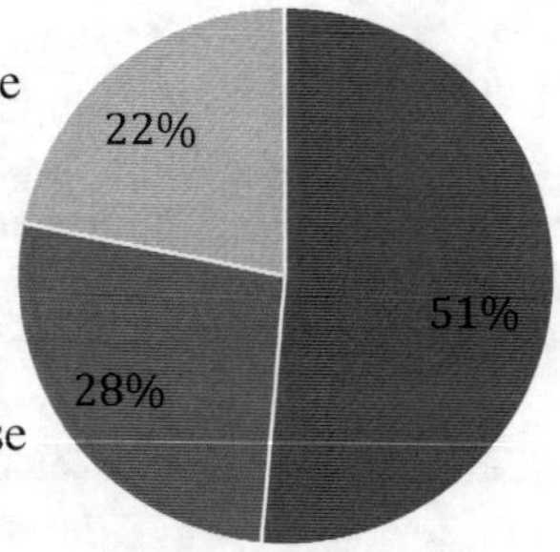

Figure 2. Adaptive reuse methods

6. Conclusion

Eventually, Preservation approaches become the most popular approach in adaptive reuse due to its successes in distinguishing between original and contemporary. Rather than blurring old with new, and causing confusion between historic and contemporary, these elements accentuate each other. Furthermore, this method is emotionally attractive for seeing and perceiving such old structures and forms because it helps preserve the existing atmosphere of the historic districts.

7. Acknowledgment

All data collection is a collaboration effort between interior architecture from Universiti Sains Malaysia (Malaysia), and Chulalongkorn University (Thailand) in conjunction with USM-Chula Workshop held from 21 to 24 Mac 2018 in Penang. This article also expresses appreciation to Dr Zarifah Maliki from Universiti Sains Malaysia, Assist Professor Sake Sawasdee and Dr Sirintra from Department of Interior Architecture, Faculty of Architecture, Chulalongkorn University.

References

[1] Bollack, F. A. (2013). *Old buildings, new forms: new directions in architectural transformations*. United States: Monacelli Press.
[2] Brooker, G., & Stone, S. (2008). *Basics interior architecture 02: Context & environment* (Vol. 2): Ava Publishing.
[3] Bullen, P., & Love, P. (2011). A new future for the past: a model for adaptive reuse decision-making. *Built Environment Project and Asset Management,* **1**(1), 32-44.
[4] Bullen, P. A., & Love, P. E. (2009). Residential regeneration and adaptive reuse: learning from the experiences of Los Angeles. *Structural Survey,* **27**(5), 351-360.
[5] ElKerdany, D. (2002). *Innovation and conservation: case studies in intervening with valuable buildings.* Paper presented at the Proc. UIA Int. Conf. WPAHR on Architecture and Heritage as a Paradigm for Knowledge and Development Bibloliotheca Alexandrina, Egypt.
[6] Eyüce, O., & Eyüce, A. (2010). Design education for adaptive reuse. *International Journal of Architectural Research: ArchNet-IJAR,* **4**(2/3), 419-428.
[7] Kim, D., M. of Architecture. (2018). *Adaptive Reuse of Industrial Buildings for Sustainability; Analysis of Sustainability and Social Values of Industrial Facades.* (Thisis), Retrieved from http://hdl.handle.net/2152/67980
[8] Mustafa, N. A., & Abdullah, N. C. (2013). *Preservation of cultural heritage in Malaysia: An insight of the National Heritage Act 2005.* Paper presented at the Proceeding of International conference on tourism development: Building the future of tourism.

[9] Ozik, D. (2006). *Reinvention through reuse: strategies for the adaptive reuse of large-scale buildings.* Massachusetts Institute of Technology., Massachusetts Institute of Technology. Retrieved from http://hdl.handle.net/1721.1/37578

[10] Pimonsathean, Y. (2002). Current issues concerning adaptive re-use in the conservation of urban cultural heritage. *RC: Revista de Cultura= Review of Culture,* **4**, 45-51.

[11] Plevoets, B., & Van Cleempoel, K. (2011). Adaptive reuse as a strategy towards conservation of cultural heritage: a literature review. *Structural Studies, Repairs and Maintenance of Heritage Architecture XII,* **118**(12), 155-163.

[12] Plevoets, B., & Van Cleempoel, K. (2014). Aemulatio and the interior approach of adaptive reuse. *Interiors,* **5**(1), 71-88.

[13] Reuse, D. A. (2004). Preserving our Past, Building our Future. *DEPARTMENT OF ENVIRONMENT AND HERITAGE. ACT: Department of Environment and Heritage, Commonwealth of Australia,* **17**.

[14] Yildirim, M. (2012). Assessment of the decision-making process for re-use of a historical asset: The example of Diyarbakir Hasan Pasha Khan, Turkey. *Journal of cultural heritage,* **13**(4), 379-388.

IOP Conf. Series: Materials Science and Engineering **636** (2019) 012015 doi:10.1088/1757-899X/636/1/012015

Adoption of Social Media by Architecture Students in Fostering Community Service Initiative using Technology Acceptance Model

Dr. Sujatavani Gunasagaran [1], Dr. M Tamilsalvi Mari[2], Dr. Sucharita Srirangam[3] and Dr. Sivaraman Kuppusamy[4]
[1, 2, 3]School of Architecture, Building and Design
Taylor's University, Selangor, Malaysia
[4]School of Built Environment, University of Reading Malaysia,
Iskandar Puteri, Johor, Malaysia.

Sujatavani.Gunasagaran@taylors.edu.my

Abstract. The use of the Internet for social networking is very popular amongst the youth. Collaborative technology leads to instant online community that enables fast and convenient communication. Adoption of Social Media has been researched on many contexts. However, little research has paid close attention to the student's acceptance of Social Media (SM) as e-learning in higher educational teaching, especially in architectural education. This study aims to investigate factors that affect acceptance and behaviour of architecture undergraduate students towards the usage of SM. A quantitative method was used and qualitative responses of 33 architecture students using social media in their elective module, Community Service Initiative (CSI), was analyzed using the Technology Acceptance Model (TAM). TAM model for this study focused on social norm (SN) as an external factor, the two user beliefs, perceived usefulness (PU) and perceived ease of use (PEU) and behavioural intention to use (BI). Factor Analysis and Pearson Correlation showed significant relationship between social norm, perceived ease of use, perceived usefulness, and behavioural intention to use. The results of this study provided evidence for the potential use of social media as e-learning in similar modules.

Keywords. *social media; e-learning; technology acceptance model (TAM); community service; architectural education*

1. Introduction

E-learning has been introduced and adopted rapidly by higher educational institutions to facilitate the teaching and learning of students. Despite the potential of e-learning as a tool to enhance education and training performance, its value will not be realized if users do not accept it as a learning tool [9]. The Technology Acceptance Model (TAM) is a concept model that was most widely used in measuring technology or e-learning acceptance. The model used in this study adheres to the original structure proposed by Davis in 1986 [5]. In this study, the social norm (SN) or influence of instructors and peers on forming the perceived usefulness (PU) is used. Facilitating conditions, the external factor

Content from this work may be used under the terms of the Creative Commons Attribution 3.0 licence. Any further distribution of this work must maintain attribution to the author(s) and the title of the work, journal citation and DOI.
Published under licence by IOP Publishing Ltd

for perceived ease of use (PEU), was not used as the students have prior experience of using all the social media and have access to facilities such as computers, mobile devices, and internet. PEU refers to the degree to which an individual expects no physical and mental difficulties in adopting the technology [4]. Most TAM studies that consider technology adoption, predicted actual usage by measuring only behavioural intention to use the system [15]. Behavioural intention in TAM was also presumably formed as a result of careful decision-making process [16]. In this research, the behavioural intention is assumed as predictor of actual use. This study aims to investigate the factors that influence behavioural intention to use social media (SM) in fostering community service learning among architecture students.

The module Community Service Initiative (CSI) is an elective taken by architecture students in a private higher educational institution in Malaysia. Community service is a form of service learning, which uses curriculum to foster a sense of care and concern among students for their community and the world at large. This module aims at recognizing and taking steps to meet the needs of others. Students were exposed to community needs and experience caring for others while learning about the resources available to them for revitalizing their communities. This gives them an opportunity to use and enhance their creative, critical, and reflective skills. Architecture students are expected to have all these skills as they are exposed to them since semester 1 year 1, especially in design studio teaching. Thus this study analyses the architecture students' acceptance of SM in fostering community service initiative's learning outcome.

There are wide usages of SM in promoting e-learning in higher education. The normal usage of social media in e-learning is through formal and informal learning. Formal learning is a highly structured type of learning that happens in courses, classrooms, and schools, resulting in learners receiving grades, degrees, diplomas, and certificates. Whereas informal learning rests primarily in the hands of the learner and happens through observation, trial and error, asking for help, conversing with others, listening to stories, and reflecting on a day's events, or its stimulated by general interests [2][13]. Facebook was categorized as a social networking site and YouTube as creativity works sharing site [10]. The usage of SM for teaching and learning is categorized as the constructivist approach to learning as in Figure 1 [2]. The researchers found that SM is useful in teaching and learning especially in this elective module, community service initiative, thus, the need to find and understand the student's acceptance of it.

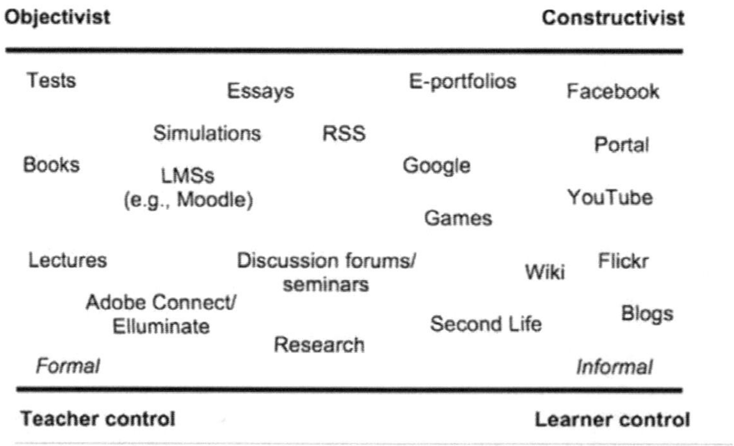

Figure 1: Analysis of web 2.0 tools, an educational perspective [2].

Constructivism suggests that the learner is actively involved in learning. The constructivist learning theory and how it describes the way that learners construct meaning through our interactions with one another, the community, and the environment, and that knowledge is not something absolute [7]. Thus, in this module, Community Service Initiative, students learn or create (construct) their own meaning by interacting with instructors, peers, public, and community in need. In a study conducted to study the use of digital simulation tool among architecture students, the finding was that that architecture students were able to adopt and use new knowledge and skills followed by analysing skills that are important for critical thinking [14]. In this study Facebook and YouTube were suggested by the instructor in an attempt to reach out to a larger audience than the campus community. The students also used WhatsApp, an instant messaging smartphone application and others. It is expected that architecture students' community service initiative will reach community at large with healthy interaction and positive outcomes.

The effective use of information and communication technologies in delivering e-learning based components of a course is of critical importance to the success and student acceptance of e-learning [11]. The instructor found that using the SM is very relevant to informal learning and the various activity that the students might choose to accomplish their aims such as create awareness, to promote a cause and maybe raise funds. Using the SM, the students will be able to reach out to more people other than the campus community. The students chose disabled children's home as their choice of community in need and planned activity to foster communication and understand problems related to this community. Then the students are to suggest activities to help them. The students in groups chose to create social awareness by creating Facebook pages, organizing an exhibition, and selling merchandises such as key chains and t-shirts to raise funds. They used SM extensively for this purpose. It is also a requirement for the architecture students to reflect and document their projects in a blog called e-portfolio.

The learning outcomes prescribed for this CSI elective module are;

1) Develop relevant communication skills for collaborative work with community/society and/or to work in a career that will contribute to society.
2) Develop a holistic awareness of the variety of problems (e.g. the environment, poverty, and social discrimination) that call for action; critically analyze how such problems should be approached and how to solve them scientifically.
3) Apply information management skills and strengthen their lifelong academic learning.
4) Develop a strong sense of social responsibility and be able to portray leadership qualities and be able to work effectively as a team.

2. Method

To identify the user acceptance of SM, the data for this study was collected using survey questionnaire administered to 33 undergraduate Architecture students from the March 2016 intake. Students were instructed in the survey to offer information about their experiences of using social media. The survey adhered to the Technology Acceptance (TAM), constructs and uses a five-point Likert scale (1 for strongly disagree to 5 strongly agree). SPSS application (Version-20) was used to analyze the data. The instrument used for this study was designed based on the objectives of the study. The TAM for this study is illustrated in Figure 2. The researchers also observed the usage of SM by following the Facebook, YouTube, and WhatsApp postings.

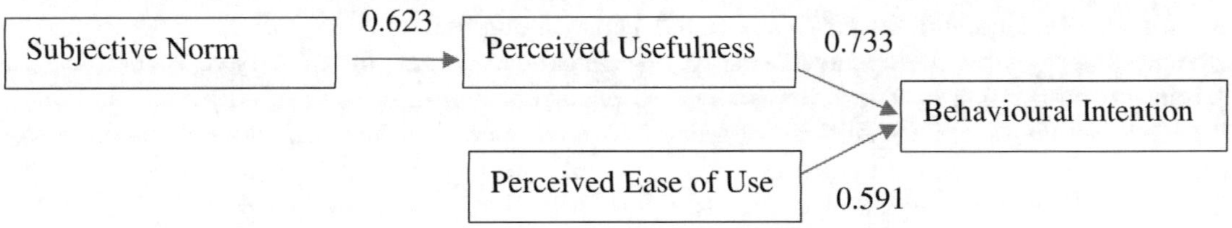

Fig. 2: Technology Acceptance Model for this study.

The participants in this study include 54.5% male and 39.4% female students. They come from various educational experiences prior to this undergraduate architecture programme. 87.9% of the students were in semester 5, final year, and 12.1% of the students from sem 3 and 4 from year 2. All students have prior experience of using social media (SM). 69.7% reported using social media for 6 years or more, 27.3% of students used the SM for 3 to 4 years, and one student had used for 1 to 2 years.

3. Results and Discussion

Subjective norm affects the general form of a student's personal perception of social media (SM) usage, particularly as a degree of perception to which an individual perceives how important others believe he or she should use the SM in achieving the learning outcome. In the higher educational institutions, perception of usefulness will be formed by the social influence of instructors and peers. The students, then, would have a positive attitude towards the use of SM when the instructors ascertain that the SM is useful to them. It is noticed that subjective norm and perceived usefulness has a strong positively correlation, from table 1, $r = 0.623$, $p \leq 0.05$. TAM has been extensively used and extended to study in an e-learning context, where, the constant use of the two use beliefs in TAM, perceived usefulness (PU) and perceived ease of use (PEU), were used to explain university students' acceptance of e- learning. In this study, the strongest correlation was found between perceived usefulness and behavioural intention to use, $r = 0.733$, $p \leq 0.05$. The architecture students strongly feel that SM is useful in achieving their aim. In a SM usage study it was explained that Facebook is user driven and has the ease of uploading a video, a picture, or a message going viral [12], thus, the importance of students choosing this social media to achieve their learning outcome for this module. The students also have prior experience of using the social media with 69.7% of the student using it more than six years.

Table 1 *Pearson's Correlations Table*

	PEU	PU	SN	BI
PEU	1			
PU	0.546**	1		
SN	0.300	0.623**	1	
BI	0.591**	0.733**	0.413*	1

**. Correlation is significant at the 0.01 level (2-tailed).
*. Correlation is significant at the 0.05 level (2-tailed).

Most students used and intended to use Facebook as their chosen social media platform for their studies. The second social media that was used and was intended to be used for study was WhatsApp, an instant messaging smartphone application that allows sharing of digital media such as images and videos instantly. This WhatsApp application is more private as it needs the phone number of the user, while Facebook allows user to be available for public, thus, having more audiences and may attract participation of the community service initiative cause highlighted.

According to table 1, perceived ease of use and behavioural intention have moderate positive relationship, $r = 0.591$, $p \leq 0.05$. This finding is concurrent with Davis [4] in which it was found that perceived usefulness has a significantly greater correlation with usage behaviour than did perceived ease of use. Perceived ease of use had a significant effect on attitude towards usage [8]. 52% of architecture students use social media for studying more than 6 hours daily, while 33% use it for 3-4 hours daily and 15% use it for 1-2 hours. It can be concluded that students use SM with ease with prior experience and daily usage. The researchers' observed that students tend to use SM for

IOP Conf. Series: Materials Science and Engineering **636** (2019) 012015 doi:10.1088/1757-899X/636/1/012015

communication purposes within group mates and community involved to manage tasks, and as an archive to share materials such as pictures and videos rather than for campaign usage or for communicating with general SM users. In order to understand the students' perception of the usefulness of social media (SM) in achieving the CSI elective module, the survey also had a question; "Without SM, will you be able to achieve your aim?" To this question, 67% of the students answered that without SM the aim will not be achieved which is consistent with their belief that SM is useful.

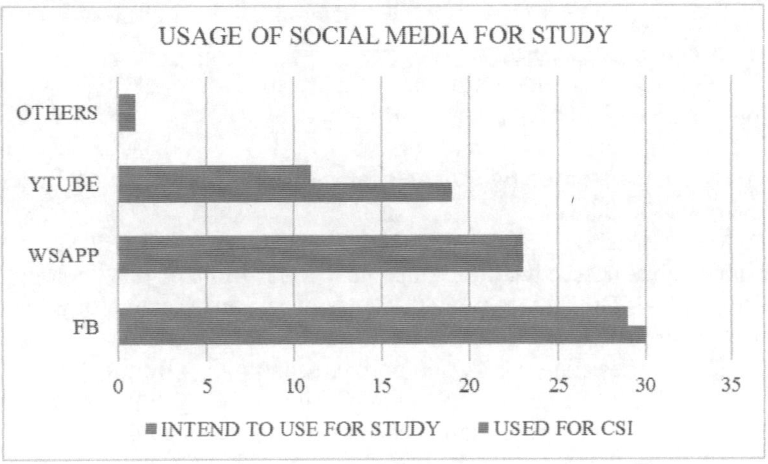

Fig. 2: Graph showing the number of students that have used SM for this CSI elective module and will use the SM in their future studies.

4. Conclusion

This study uses TAM framework for enhancing the understanding of students' acceptance of using SM in fostering community service initiative's learning outcome. The finding suggests that perceived usefulness had a greater impact on behavioural intention than perceived ease of use. It can be concluded that students use SM with ease with prior experience and daily usage. 67% of students also believe that SM is important in achieving the aim of this module, while the rest of the 33% might have relied on campus community to achieve their aim. Some TAM researchers argue that students may have different motivations, such as grades, rewards, and so on, [11] thus, do not rely on SM in achieving the aim of the study. The instructor opinioned that the students who used SM are more open for new challenges in learning and believed that the cause they are promoting needs a larger audience. The researchers also found that students use the SM with ease for communicating within the community involved in their project and use SM as an archive with ease. Findings of the study present important implications for educational institutions in using SM as e-learning [1], which implies that the use of SM in learning can connect students and instructors, increase learners' motivational level, and can create strong communities of practice for teaching and learning, which can expand the learning process beyond the boundaries of a traditional classroom. If usage of SM is to be considered for teaching and learning in a module, a guideline for usage of SM needs to be consolidated with clear aims to be established because the potential of SM usage is varied and would be a valuable learning experience for students. The study focused only on architecture students enrolled in a module. The findings also give confidence to the view that using SM is particularly suited to this elective module as e-learning because the architecture students are relatively experienced in using Social Media, facilities are readily available, and almost all students have a smart phone. The researchers suggest that in the future, a larger sample with respondents representing different modules can be used to generalize the results.

References

[1] Al-rahmi, W., & Othman, M. (2013). Using TAM model to measure the use of social media for collaborative learning. *International Journal of Engineering Trends and Technology (IJETT)*, *5*(2), 90-95.

[2] Bates, T. (2011). Understanding Web 2.0 and its implications for e-learning. In Web (Vol. 2, pp. 21-41).

[3] Dabbagh, N., & Kitsantas, A. (2012). Personal Learning Environments, social media, and self-regulated learning: A natural formula for connecting formal and informal learning. The Internet and higher education, 15(1), 3-8.

[4] Davis, F. D. (1989). Perceived usefulness, perceived ease of use, and user acceptance of information technology. MIS quarterly, 319-340.

[5] Davis, F. D., Bagozzi, R. P., & Warshaw. P. R. (1989). User acceptance of computer technology: A comparison of two theoretical models. Management Science, 35(8), 982-1003. doi: 10.1287/mnsc.35.8.982

[6] Dhume, S. M., Pattanshetti, M. Y., Kamble, S. S., & Prasad, T. (2012, January). Adoption of social media by business education students: Application of Technology Acceptance Model (TAM). In 2012 IEEE International Conference on Technology Enhanced Education (ICTEE) (pp. 1-10). IEEE.

[7] Harasim, L. (2012). Learning theory and online technologies. Routledge.

[8] Hu, P.J., Chau, P.Y.K., Sheng, O.R.L. and Tam, K.Y. (1999) 'Examining the technology acceptance model using physician acceptance of telemedicine technology', Journal of Management Information Systems, Vol. 16, No. 2, pp.91–112.

[9] Lee, B. C., Yoon, J. O., & Lee, I. (2009). Learners' acceptance of e-learning in South Korea: Theories and results. Computers & Education, 53(4), 1320-1329

[10] Mangold, W. G., & Faulds, D. J. (2009). Social media: The new hybrid element of the promotion mix. Business horizons, 52(4), 357-365.

[11] Ratna, P. A., & Mehra, S. (2015). Exploring the acceptance for e–learning using technology acceptance model among university students in India. International Journal of Process Management and Benchmarking, 5(2), 194-210.

[12] Rauniar, R., Rawski, G., Yang, J., & Johnson, B. (2014). Technology acceptance model (TAM) and social media usage: an empirical study on Facebook. Journal of Enterprise Information Management, 27(1), 6-30.

[13] Selwyn, N. (2007). The use of computer technology in university teaching and learning: a critical perspective. Journal of computer assisted learning, 23(2), 83-94.

[14] Sujatavani. G., & Mari, M. T. (2015). Using Digital Simulation as an e-learning tool to create dynamic learning in Architecture students. Global Journal of Business and Social Science Review. 3(1), 107-114.

[15] Turner, M., Kitchenham, B., Brereton, P., Charters, S., & Budgen, D. (2010). Does the technology acceptance model predict actual use? A systematic literature review. Information and Software Technology, 52(5), 463-479.

[16] Venkatesh, V., Morris, M.G., Davis, F.D. and Davis, G.B. (2003) 'User acceptance of information technology: toward a unified view', MIS Quarterly, Vol. 27, No. 3, pp.425–478.

Typologies of design thinking: the learner-led perspectives

S Srirangam[1,5], V F P Ng[2], T Mari[3], S Gunasagaran[3] and S Kuppusamy[4]

[1] Senior Lecturer, School of Architecture, Building and Design, Taylor's University, Selangor, Malaysia. sucharita.srirangam@taylors.edu.my

[2] Associate Professor, School of Architecture, Building and Design, Taylor's University, Selangor, Malaysia.

[3] Senior Lecturer, School of Architecture, Building and Design, Taylor's University, Selangor, Malaysia. Malaysia.

[4] School of Built Environment, University of Reading Malaysia, Iskandar Puteri, Johor, Malaysia.

[5] Author to whom any correspondence should be addressed.

Abstract. The research aims to offer critical discourse on various perspectives on design studio teaching pedagogy. The paper begins by exploring the generic literature that focuses on a range of instruments, approaches and actions; the tools that the designers deploy in order to steer specific thought process in design. By keeping the learners at the core, the subjectivity of designers brought to main focus here. Although there are many research works on studio environment and studio framework, only a very few focus on capturing evidences of the learners' creative thinking and this research is an attempt to fill this gap and keeps the learners at the core of analysis. The method was a theoretical exploration on various perspectives on design thinking and to obtain evidences of students' thinking. Two studios of Final Year studies of Part I, from University Malaya and Taylor's University, have been the case studies of the research. The paper first offers a critical discourse on identifying a gap in architectural design studios' teaching pedagogy and second offers a set of typologies of design thinking from the case studies. The findings were 8 typologies of design thinking which could be categorised into 3 groups, based the domain of the constructs. The research reinforces and determines the importance of learner-led teaching pedagogy for architecture design studios, by formulating typologies of design thinking.

Keywords: architectural education, design thinking, student-centred learning, typologies

1. Introduction

The research has its objective to offer critical discourse on various perspectives on design studio teaching pedagogy. The paper begins by exploring the generic literature that focuses on a range of instruments, approaches and actions, the tools that the designers deploy in order to steer specific

Content from this work may be used under the terms of the Creative Commons Attribution 3.0 licence. Any further distribution of this work must maintain attribution to the author(s) and the title of the work, journal citation and DOI.
Published under licence by IOP Publishing Ltd

thought process in design. By keeping the learners at the core, the subjectivity of designers brought to main focus here. This research takes a typological exploration of design thinking through evidence based research. The usefulness of such a research are a) to promote individualism and originality of design thinking, b) to offer to effective teaching pedagogy to empower the inner strength of the young designer/learner and c) to offer to the idea of learner centered teaching. The major research question is: What is the typology of thinking processes of a student in a studio environment?

The sub research questions are:

a) what are the various stances of design thinking in the existing literature?
b) are there any typologies of design thinking in a studio environment?

The methods were to explore theoretically on various perspectives on design thinking through literature review and to obtain evidences of students design thinking, through two design studios in University Malaya and Taylor's University. Whilst the former method answers the sub research questions a above; the latter captures evidence for the sub research question b, above.

2. Theoretical Discourses

2.1. *Existing literature on design thinking*

By so far the exploration and research done for and in the domain of design thinking fall into three main themes, namely, a) studio resources, b) teaching pedagogy and c) mediums of productivity. Firstly, the theme of studio resources considers the studio ambience and instructors. Ochsner (2000) stresses the interaction between the student and the instructor as crucial and identifies the 'shared play' as telling and listening and demonstrating and imitating. The findings of Attoe and Mugerauer (1991) through interviewing 20 award winning studio masters, effective studio teaching comprises of 14 factors parked under three categories, namely, the teacher as the self (4 categories), personal style (4 factors), course format (6 factors). Secondly, the theme on teaching pedagogy elaborates on the processes involved in design activity. Lawson (2005), mentions the stages of creative process (first sight, preparation, incubation, illumination and verification) and the types and styles of design thinking (behaviorist, the Gestalt school, cognitive science approach) considering the reasoning and imagining as important to designers. All these involve in and engage with the orderliness of thought processes. Carmona (2017) focuses on the tools of governance of design process, such as the context and the self. Rodgers & Winton describe (2010) processes such as inspiration, ideation and implementation; Ardington and Drury (2017) explore the direction on studio as context of learning and the discourse of guidance and feedback; Atlier project (2011) suggests designing as performing in which performance contributes interventionist, participative and experiential epistemology. Brown (2009) mentions design is act of changing a need to a demand or opportunity; stresses on divergent ways of looking into new choices and convergent ways for making that selected choices as an opportunity and possibility. Finally the medium of productivity such as drawings (Unwin, 2007), models and metaphors (Snodgrass & Coyne, 2013) and digital technology (Tepavčević, 2017).

Very few research works focus on capturing evidences of the learners' creative thinking. Snodgrass and Coyne (2013) explore the deeper interpretations and elaborates on three design thinking namely, play, edification and otherness. Chance (2010) emphasized architectural education builds skill in all areas of Kolb's (1984) (as sighted in Chance, 2010) decision-making cycle as shown in Figure 1. The cycle involves active experimentation, concrete experience, reflective observation, and abstract conceptualization. Architectural education requires students to use all of these modes of thinking. Most studio assignments require at least one full cycle, and each smaller decision benefits from holistic analysis as well. Chance (2010) proving that linear thinking just doesn't work for architects using Kolb's model. Kvan and Jia (2005) used Kolb's model to explore learning styles of architectural students in China and correlates their learning styles with design studio performance. Their studies noted that assimilators succeeded by achieving higher grades than others in the design studio.

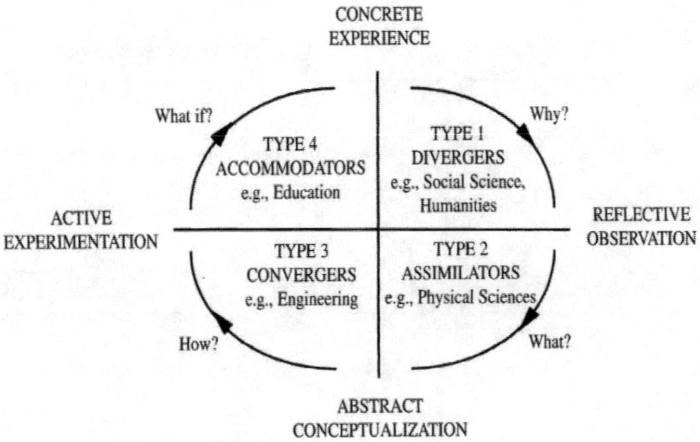

Figure 1. Learning styles and learning cycle based on Kolb's model (Chance, 2010).

The above literature review brings in the following issues:

a) The external aspects such as studio environment, teachers and context of the course work are explored in detail, whereas the designer or the 'self' is given almost least exploration.
b) The learners' background, choice of thinking, approach to architecture and possible originality of design thinking are not focused at all.
c) It is widely understood that there is a gap on exploring the connections between process (resources and methods) and product (the final outcome), which is the designer/learner herself/himself.
d) Lack of a studio method that actually empowers a student with originality of her/his own design thinking.

3. Case study

3.1. *Design Studio V (DSV), University Malaya*
Further to the pilot study conducted in Taylor's University during March to July 2018, the first case study was conducted in University Malaya, Design Studio V. This semester's design following the year's themes 'tourism + culture', the project was to propose a compact urban hotel for Kuala Terengganu. This new sub typology of accommodation that will be the first of its kind in city. The design proposal must take into account the definition of what is an urban compact hotel. The accommodation should provide a world-class compact hotel experience for visitors to Kuala Terengganu that offers a level of service and privacy within the tight space afforded by a small parcel of land. A new hotel building that will permit guests to enjoy the several unique urban garden settings, that also is be used to buffet the street noise.

The cohort of August to December June 2018, has just completed the Studio V project on the compact urban Hotel. The evidences show that the thought processes can be generalised and patterns of collectiveness are evident. Our initial preposition, of finding if there could be themes of design thinking, has proven to be positive at the outcome.

The typologies of design thinking were interesting; as below.

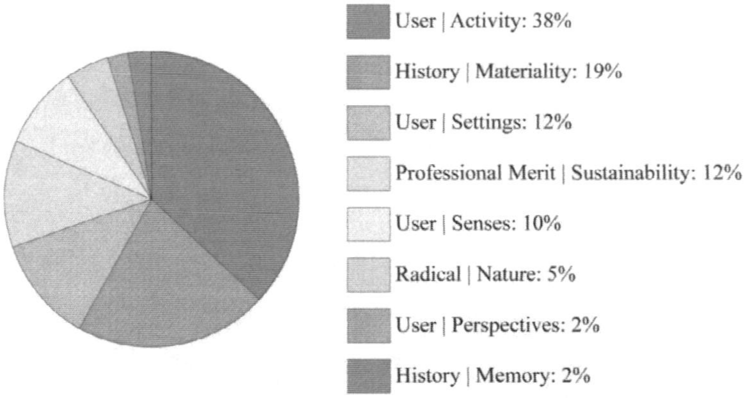

Figure 2. Typologies of design thinking: 42 studio-students from University Malaya.

The explorations within a typology varied in design intension and in consideration of design components. This can be explained as below, from the data collected:

User | Activity: The typology of 'users | activity' has had seen a building not only brings people together through culture and the arts, but it also functions as a nexus between the various streams of the city. The design considerations were: functionality, comfort (indoor climate, lighting), aesthetics (materials, shapes, details), the relationship with the context (social, cultural and physical), social facilities (spaces of interaction), fluency (ease of movement, universal design) and the environment which relate to the surroundings.

History | Materiality: The typology of 'History | Materiality' highlighted the difference between memories understood as a materiality versus a social or cultural fact. The design considerations were: Tranquillity, Social interactions, Greenery, Landscape, Eateries, Farming, History, Textile, Identity, Process, Abstraction, Sense of tradition, Structure, Technology, Story-telling, Lighting, Safety, Pattern, Local material used, Assembly, Performance, Shadow play, Angle, Movement, Human body, Façade, Historical context, construction materials, spaces, culture

Professional Merit | Sustainability: The typology of 'Professional merit | sustainability' was to create spaces in sustainable way is not only in the strict sense of implementing green technologies, but socially and culturally as well. The design considerations were Ecology, Green buffers, Public spaces, Green spaces, Open spaces, Greenery, Façade, Climatological, Ventilation, Shading, Façade pattern, Connectivity, Junctions, Safety, Traditional, Layout, Pattern, Farming, Trees, Recycle, Garden, Courtyard.

User | Settings: The typology of 'User | settings' encouraged different adaptations might include creating spaces that are green, encourage activity, or even promote social interaction. The design considerations were: image, Community, Accessibility, Engagement, Ease of recognition, Ventilation, Traditional, Layout, Pattern, Spatial arrangement, Boat craft, Spatial arrangement, Boat craft, Spatial sequence, Workspaces, Sense of place, Culture, Traditional character, Interactive, Circulation, Recreation

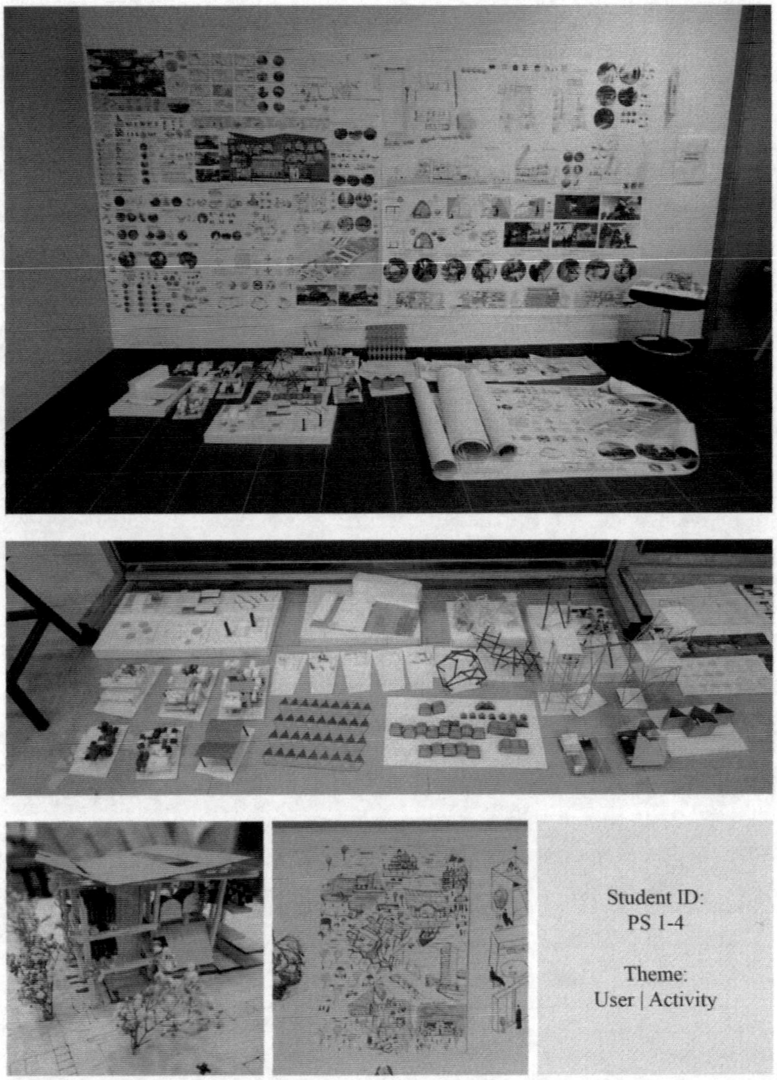

Figure 3. Sample of a University Malaya's student work under typology of User| Activity.

User ❘ Senses: The typology of 'User ❘ sense' had perceived the sources of senses such as sight, smell, hearing, and touch far above what is normal for the user's needs. Those tactile sensations are not just changing general orientation or putting people in a good mood, they have a specific tie to certain abstract meanings towards social, culture and digital media in future. The design considerations were: Visual, Adaptability, Shadow, Light, Human scale, Story-telling, Identity, Movement, Digital media, Sound, Senses, Smell, Pattern.

Radical ❘ Nature: The typology of 'Radical ❘ Nature' explained that design should be paid sufficient attention to the natural rhythms of human neurobiology can result in conscious joy. The design considerations were: Health, Nature, Art, Spatial sequence, Public spaces, Buffer zones, Connectivity, Blurring boundaries

User | Perspectives: The typology of 'User | Perspective' related to functional layout to space planning and works towards a functional layout should be more concern in safety, security and privacy which encourage social interaction. The design considerations were: Flexibility, Safety, Security, Privacy, Social interactions and views.

History | Memory: The typology of 'History | Memory' created a mutual understanding of the project and to build up ownership and roots in the local community. The design considerations were: Memorability, History, Technology, Structure.

3.2. *Design Studio V, Taylor's University*

The second case study has been conducted at Taylor's Design Studio V of the B.Sc. (Hons) in Architecture. The studio focuses on urban context with corner block site for a Learning Centre typology; comprising of architectural strategy and design development. The major site issue being the fall of spatial connections and engagement to the local community, to the tides of various contemporary forces. The design of the proposed library is to consist of appropriate architectural responses that address aspects of the urban street, its context, and surrounding user behavioural patterns as discerned and analysed in the Preliminary Studies.

Table 1. Design Studio V Pedagogical chart.

Levels of knowledge	*Pedagogy*	*Stage*	*Methods*
Declarative	Collaborative	Analysis and research	Preliminary studies: Group project, Online discourses, precedent and urban studies
Procedural	PBL	Design and evaluation	Final project: Individual design explorations, design and design development

The cohort of August to December 2018, has just completed the Studio V project on a learning Centre. Students' domains of design thinking are captured through their process of work and the product of final design outcome at the stage of design development and detailing. During the project, the students were observed and provided with guidance that lignified their own opinions and ideas. The evidences show that the thought processes can be generalised and patterns of collectiveness are evident. Our initial preposition, of finding if there could be themes of design thinking, has proven to be positive at the outcome.

The typologies of design thinking were interesting; as below.

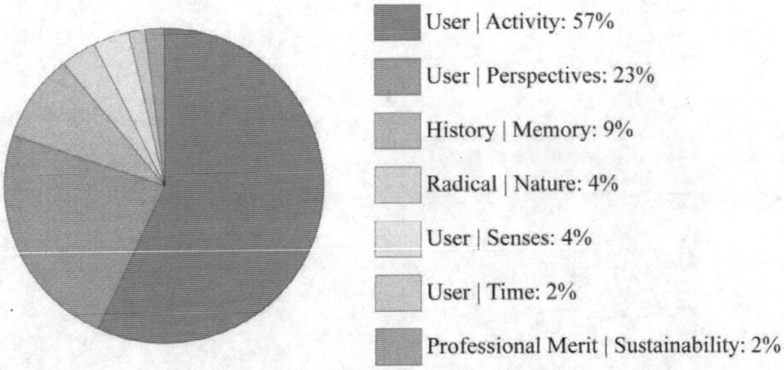

Figure 4. Typologies of design thinking: 56 studio-students from Taylor's University.

The explorations within a typology varied in design intension and in consideration of design components. This can be explained as below, from the data collected:

User | Activity: The typology of 'user | activity' was more concerned of everyday public activities to the periodic maintenance requirements as well as how the elements, zones and spaces are organised. There are different techniques students often use to explore and explain programme. The design considerations were: Community, functions, Social interaction, Adaptability, Story-telling, City image, Social experiences, Readability, Historical value, Privacy, Safety, Security, Human needs, Street connectivity, Street, Pedestrian. Human needs.

User | Perspectives: The typology of 'User | Perspective' establishes a link between the user and community, earning their trust, admiration as well as enhancing physical comfort and safety and security of neighbourhood. The design considerations were: Visual aspects, comfort, Sense of belonging, Places for exploration, Local attachment, Experiences, Social interaction, Movement, Human scale, Walkability, Privacy, Positive space, Greenery, Expression, Dynamism, Pattern, blurred boundaries, Building edges, Landmark, Permeability, , Circulation, Pedestrian, Connectivity, Network, Perception, Aesthetic, functions.

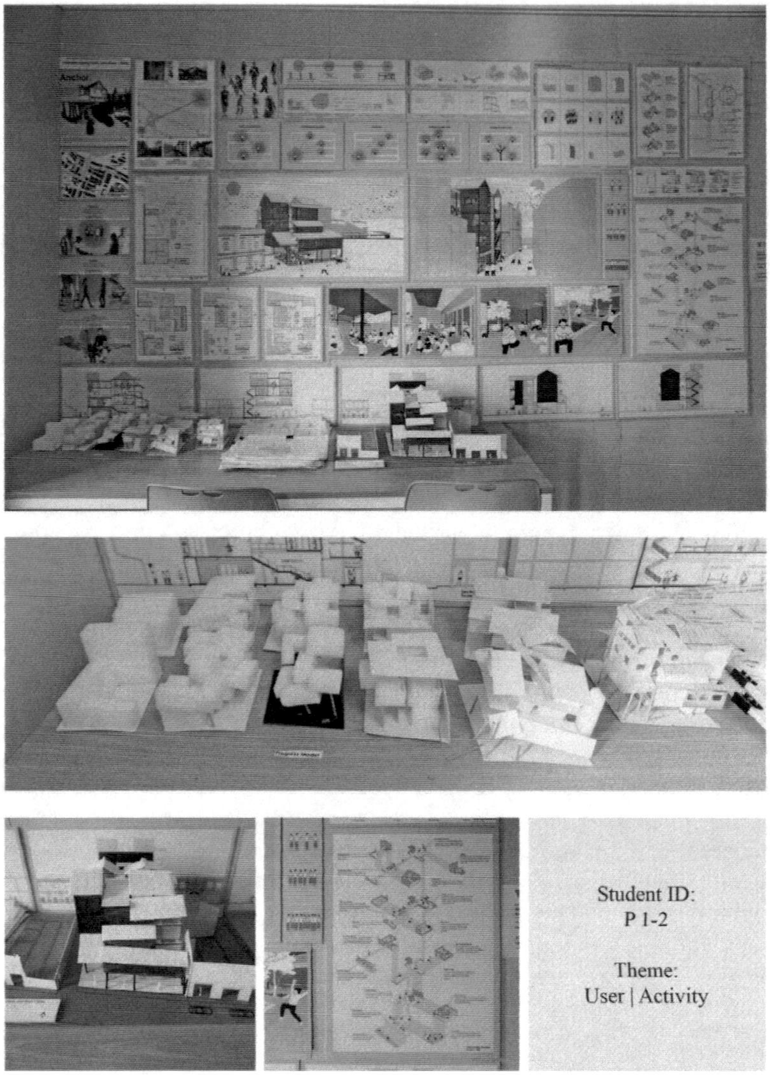

Figure 5. Sample of a Taylor's University student work under typology of User| Activity.

History | Memory: The typology of 'History | memory' refers to spaces that can be remembered, that we can imagine or hold in the mind as well as consideration. The design considerations were Cultural value, Healing, Social belonging, Memorability, Traditional, Experiences, Vernacular, Memory, History, Art, Craft, Identity, Community, Collaboration.

Radical | Nature: The typology of 'Radical | nature' offers the physical nature of the built environment an easier way to assume that the visual qualities of spaces dominate our perception of a building. The design considerations were Digital media, Nature, Technology, Interactive, Tradition, Greenery, Landscapes, Squares, Public spaces, Perception.

Professional Merit | Sustainability: The typology of 'Professional Merit | Sustainability' ensuring health lives and promoting well-being for all is important to building prosperous societies. Sustainability challenges the professional to contribute to sustainable communities and quality of life.

The design considerations were: Cycling, Environmental, Ventilation, Quality, Façade, Planting, Community

User | Senses: The typology of 'User | senses' provides an art form beyond its materiality, the less it is constrained by the natural world and the closest it is to pure spirit, the more elevated and transcendent it became. The design considerations were: Senses, Play, Music, Noise, Echo, Experiences.

User | Time: The typology of 'user | Time' concerned the buildings that develop very complex relationships to times. They are changed and even converted to other building types, which cuts them off from their networks even though they still occupy the same location. The design considerations were: Light, Shadow, Overlapping, Daylight, Active learning, day/night, weekday/weekend patterns.

3.3. *Design Studio V students from both case study universities*

In order to find constructs, which are deeper values of a student designer to engage with that particular typology of design thinking, a questionnaire survey was executed. It was online questionnaire survey open to the DSV students of both of the case study universities. The questions were open ended, in order to be led by the respondents rather than our design findings by so far in the project.

The findings from the Online Questionnaire Survey were surprising. Main finding was the initial preposition of finding whether there could be themes of design thinking or not, has proven to be positive at the outcome in the case studies chosen. The Typologies of Design Thinking as follows:

History | Materiality
History | Memory
User | Activity
User | Settings
Professional Merit | Sustainability
User | Perspectives
Radical | Nature
User | Economy
User | Senses

Secondly, it is also found that there are patterns between constructs and typologies (see Figure 18). The highest relevance of varied construct was evident when the typologies are about users – on activity (scenographics), settings (behavioural) and perspectives (formal/visual). The wealth of constructs is deep or a few when the typologies are about economic, sustainability, temporal and senses.

Thirdly, the significance is to offer tools for exploring originality in design and therefore to empower the young budding designers, the students of architecture. This finding will shed new light on learners' centred studio pedagogy with an intent of empowering the young learners. It has also offered a deep discourse on a) positivistic (materiality, settings, activity), b) interpretative (perspective, senses, time) and c) critical (sustainability, economy) thinking on the process of design itself.

4. Overall Conclusion

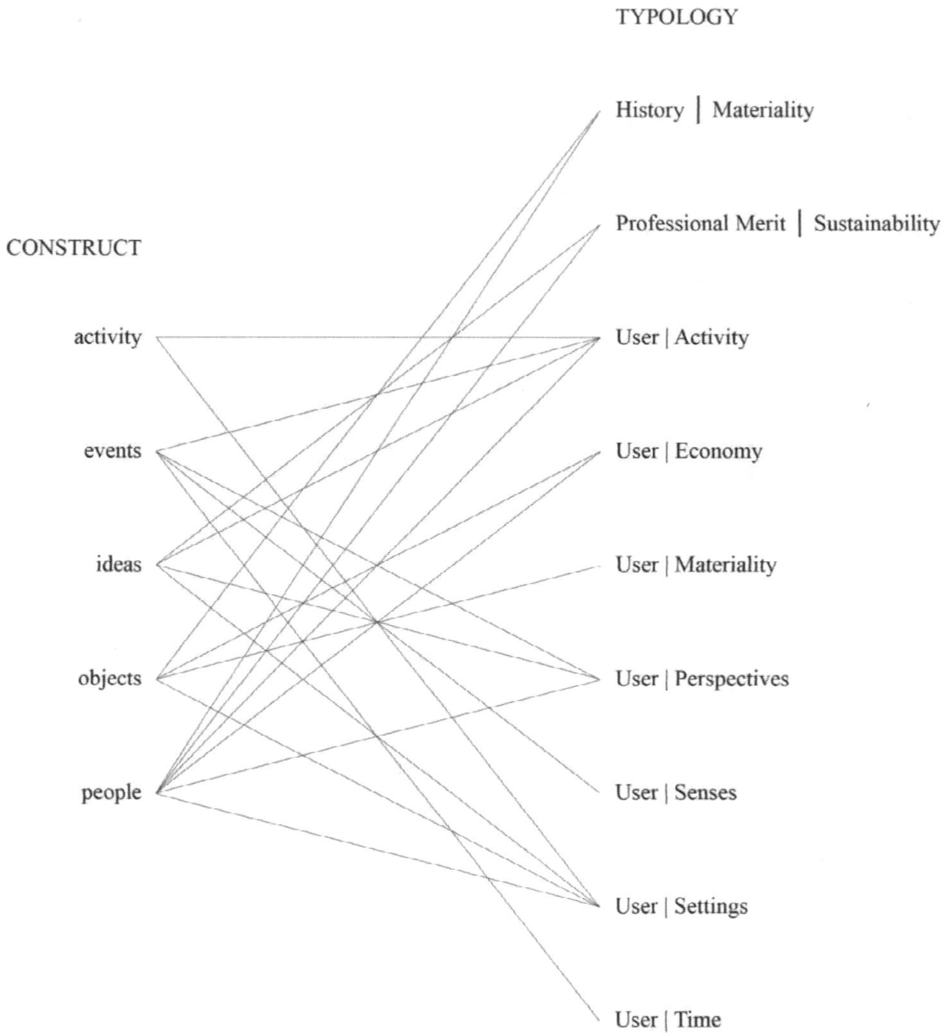

Figure 6. Constructs and typologies mapping.

Hard typologies: These formulate the typologies pertinent to the elements of design such as form, function, circulation. The constructs are usually about objects and ideas.

Soft typologies: These formulate the typologies pertinent to the intangible aspects of design such as concepts, principles and qualities. The constructs are usually about people.

In-between or the hard/soft typologies: These formulate the typologies pertinent to the middle ground of both extremes, such as the articulation of things to form a concept, scale, proportion etc. The constructs are usually about activity and events.

The future directions may be of two folds: a) recognising the field of psychology and b) pedagogy for design studios. The deeper co-relations of the constructs found in this research, leading to a specific typology of design thinking, can be executed in the field of psychology. This could enrich architecture teaching pedagogy and the future scholars seeding in the primary or secondary schools.

The research began with an intension of understanding design thinking by the young minds. It was interesting and useful to know that there are typologies of design thinking and thus the given set of students can be taught in their own unique way, through reference materials, authors and inspirations. Typologies are generic principles that will help aid a pedagogy in the studios, without disregarding the specificity of each student designer.

5. Acknowledgement

The students of the cohort of July to December 2019, of Design Studio V, of the B.Sc. (Hons) in Architecture, Taylor's University. The Studio is coordinated by Dr Sucharita Srirangam and the instructors are Ar. Patmaselvi Paramarajah, Indrani Vanniasingham, Ar. Hanani Zain, Ar. Alice Lim, Mr. Hafiz Amirol , Ar. Kenny Chong, Ar Lam Shen Fei, Mr. Emmanual Canlas, Ar Fadzwin Hashim, Mr Caleb Chow Juan, Ar Mohamad Zafarullah and Ar Ari Methi.

The students of the cohort of August to December 2019, of Design Studio V, of the B.Sc. Architecture, Architecture Department, Built Environment Faculty University of Malaya. The Studio is coordinated by Ms Ati Rosemary Mohd Ariffin and the instructors are Dr. Datin Norhayati Hussain, Mdm Norizan Abd Rahim, Dr Norafida Ab Ghafar, Ar Eddie Choi and Mr. Hiroyuki Sube.

References

[1] Ardington, A., & Drury, H. (2017). Design studio discourse in architecture in Australia: The role of formative feedback in assessment. *Art, Design & Communication in Higher Education*, *16*(2), 157-170.Atlier Project, 2011, 'Design things', Chapter 6: Designing as performing, MIT Press
[2] Atlier Project, 2011, 'Design things', Chapter 6: Designing as performing, MIT Press
[3] Attoe, W., & Mugerauer, R. (1991). Excellent studio teaching in architecture. *Studies in Higher Education*, *16*(1), 41-50.
[4] Brown, T. (2009). Change by Design, Harper Business.
[5] Carmona, M. (2017). The formal and informal tools of design governance. *Journal of Urban Design*, *22*(1), 1-36.
[6] Chance, S. (2010). Writing Architecture: The Role Of Process Journals In Architectural Education.
[7] Kvan, T., & Jia, Y. (2005). Students' learning styles and their correlation with performance in architectural design studio. Design Studies, 26(1), 19-34.
[8] Lawson, Bryan (2005), How Designers Think – the design process demystified, Architectural Press, Oxford.
[9] Ochsner, J. K. (2000). Behind the mask: a psychoanalytic perspective on interaction in the design studio. *Journal of Architectural Education*, *53*(4), 194-206.
[10] Rodgers, P. & Winton, E. (2010). Design Thinking, Conference proceedings on Engineering and Product design, Norwegian University, Norway
[11] Snodgrass, A., & Coyne, R. (2013). *Interpretation in architecture: design as way of thinking*. Routledge.
[12] Tepavčević, B. (2017). Design thinking models for architectural education. *The Journal of Public Space*, *2*(3), 67-72.
[13] Unwin, S. (2007), Analysing architecture through drawing, Building Research & Information, 35:1, 101-110

Creating Secured Residential Places: Conflicting Design Elements of Natural Surveillance, Access Control and Territoriality

M Seifi[1], A Abdullah[2], S Haron[3], A Salman[4], and M Seifi[5]

[1] Ph.D Candidate, School of Housing, Building and Planning at Universiti Sains Malaysia
[2] Professor, School of Housing, Building and Planning at Universiti Sains Malaysia
[3] Senior lecturer, School of Housing, Building and Planning at Universiti Sains Malaysia
[4] Ph.D candidate, School of Housing, Building and Planning at Universiti Sains Malaysia
[5] Msc student, School of Professional Studies, Garden City University, Bangalore, India.

E-mail: seifi.molood@gmail.com

Abstract. Territoriality, natural surveillance and accesses control are the preeminent strategies of most place-based crime prevention theories and practices. They have been extensively applied as non-detachable part of defensible space and Crime Prevention Through Environmental Design (CPTED). Even though, a large body of knowledge confirms the effectiveness of these three strategies in preventing residential burglaries, yet, they frequently diminish each other's effect. Besides, there are divergent views about their implementation, which often brings about contradictory designs and planning decisions, all of which have left architects and planners with a state of ambiguity on hiring felicitous physical elements which prevent burglaries. Hence, the current paper aims to find the design elements of the CPTED strategies which are in conflict in creating security. The study employs the qualitative method of systematically reviewing and critically analysing the literature on the subject. Conjointly, the physical features and design elements of natural surveillance, access control, and territoriality are extracted from the evidence in the literature. The results of the analysis show that the design elements of natural surveillance and access control are more recognized and employed since they are easier to understand as compared to territoriality. Furthermore, it was found that some of the physical elements of access control territoriality conflict with surveillance design elements or diminish the effects of natural surveillance on burglary prevention. Conversely, the social aspects of territoriality design strengthen the effects of natural surveillance. Therefore, the study concludes that the burglary prevention design should consider all the strategies simultaneously. Hence, it is recommended to combine the physical aspects of natural surveillance and access control with the social aspects of territoriality for successful burglary prevention. This study contributes to the body of knowledge related to CPED by shedding light on the deployment of its main strategies without conflict in their effects. However, there is a need for furthered evidence of the design elements of natural surveillance, access control, and territoriality. Only a thorough examination and comparison of the impacts of their design elements on a residential burglary in a single context could further confirm the literature findings of the present study.

Introduction

Surveillance, access control, and territoriality are the non-detachable components of the crime prevention theories such as 'eyes on the street' [1] 'defensible space' [2], CPTED [3], and other situational crime prevention practices [4]. These strategies are known to discourage crime and urge perceived safety and security against incivilities [5]. Two design elements of

Content from this work may be used under the terms of the Creative Commons Attribution 3.0 licence. Any further distribution of this work must maintain attribution to the author(s) and the title of the work, journal citation and DOI.
Published under licence by IOP Publishing Ltd

IOP Conf. Series: Materials Science and Engineering **636** (2019) 012017 doi:10.1088/1757-899X/636/1/012017

defensible space related to territoriality and natural surveillance are the demarcation of areas, which are under the influence of inhabitants to create territorial attitudes and the arrangement of buildings' windows to allow natural surveillance [6, 7]. Likewise, the physical environment level of situational measures of crime places is mainly based on territoriality, surveillance, and pedestrian and traffic flow [7]. In parallel with other theories, CPTED which was formed based on the idea of defensible space [8] consists of natural crime prevention techniques, which are achieved as a "by-product of the normal and routine use of the environment". Wherein, natural surveillance and natural access control which are mutually supportive of each other are known as the primary concepts and territoriality as the complimentary design concept of CPTED [9].

Although, natural surveillance and territoriality have been extensively applied in environmental criminology theories and burglary prevention practices, there are different perspectives about their deployment in the design of residential places. Followers of defensible space concept with 'territoriality dominant' perspective advocate that stronger territorial areas with a limited number of outsiders are less vulnerable to burglary. Newman In his book, creating defensible spaces [10] stated that an increase in a number of the non-legitimate users of the environment reduces the sense of territoriality, which ultimately leads to lower security and higher number of crime. On the same basis, Newman [11], Poyner [12], and Poyner Wedd [13] recommended restricting pedestrian movements through residential neighborhoods through the demarcation of the spaces. 'Dfensible space' design techniques encourage more closed space configuration [14]. On the contrary, the followers of 'eyes on the street' concept by Jacobs [1] with surveillance dominant perspective believe in higher admittance of people through residential spaces to prevent burglary. Jacobs also suggested placing residential blocks towards the streets to keep eyes on the streets and passers-by. Her design ideas promoted more open space configurations with emphasis on the use of surveillance elements. This controversy suggests that there could be other divergent views and elements with regards to natural surveillance, access control, and territoriality. According to [14], such a controversy is subjected to scrutiny by scholars. Hence, this paper sought to identify the spatial layouts and physical features of the environment, which have been designed to serve surveillance access control and territoriality strategies along with their influence on residential burglary. Moreover, divergent evidence in support of each strategy is compared and analysed.

Surveillance and Inter-visibility

Natural surveillance is the primary and conspicuous design concept of the physical environment for crime prevention, which facilitates observation of intruders [9]. It can be promoted through physical design that provides opportunities of watching the environment by residents and their agents as a part of capable guardianship [14]. Inter-visibility is the main component of surveillance, which is referred to the provision of self-surveillance to the legitimate users of the residential area through windows [15], visibility, and clear lines of sight without obstacles [16]. Good inter-visibility is mentioned as an inherent quality of non-burglarized houses in many Researchers' works such as [17, 18]. High degree of inter-visibility is known to be a beneficial crime prevention strategy due to the generation of the mechanisms of informal (natural) surveillance generated in the area [19]. Researchers such as [2] emphasized on strong inter-visibility of enclosed areas facilitated to only legitimate users and the arrangement of buildings' windows to allow natural surveillance of public areas. The inter-visibility within outward facing blocks in the immediate surroundings of residential places allows residents to guard strangers [20]. Furthermore, non-burglarized houses were confirmed to have greater visual contact with neighboring houses [21] and are located in areas with a high level of inter-visibility [19].

Surveillance and Accessibility

Crowe (2000), in explaining crime prevention through environmental design concepts mentioned that natural surveillance and access control components of CPTED are complementary and their design elements are not exclusive to each other [9]. Surveillance may act as the access control strategy by keeping intruders away due to an increased perception of risk [9]. Another way, the level of access control decides the level of surveillance by allowing the number of passers-by into a residential neighbourhood. The accessibility features, which promote surveillance, are the permeability, integrity, and intelligibility of the environment. With regards to permeability, the 'eyes on the street' concept of Jacobs supported the accessibility of public and flow of passers-by through streets of residential areas to guard the neighbourhood, which attracts more eyes from buildings to watch the streets [1]. She recognized the existence of a sheer number of people as the potential witnesses, whom their intervention prevents crime [22]. Researchers like [23] predicted that certain critical levels of street activity and population density were linked to a crime. A critical crime zone of intensity was, therefore, one that could support low numbers of people but insufficient densities to contain both victims and offenders. Similarly, DanzlGer In his work, "Explaining crime rate", concluded that population density influences crime rates [24]. Similarly, [25], asserted that property crimes without contact, such as residential burglary, have a negative correlation with population density. His statement was later supported by [19], which claimed that population density is shown to be inversely related to household larceny and burglary due to the improvement in visibility of crime scene to passers-by in high-density areas. In a relatively recent [25, 26] study, no correlation could be found between crime and density, however, it was confirmed that crime has a strong correlation with pedestrian movement. Moreover, [27] confirmed that chances of burglary are lower at houses with people on the street. Table1 summarizes the core findings of researches related to residential burglary; highlighting the characteristics and physical elements of surveillance in non-burglarized houses and their environment.

Table 1. Physical features and design elements of accessibility promoting natural surveillance

Attributes of secured residential places	Elements of surveillance involved in designing safe residential places	Authors	[1]
Permeable to people and bystanders for increasing witnesses and informal social control.	Residential block facing the street	Jacobs (1961)	[2]
Increased natural surveillance of inhabitants on public spaces.	Proper arrangement of windows for surveillance	Newman (1972)	[25]
Relatively populated areas	Increasing the number of people wherever needed	Shu (2009) & Angel (1968)	[19]
Increased Inter-visibility	Designing for inter-visibility	Decker et al. (1982)	[21]

Accessibility/More passers-by	Through streets instead of Cul-de-sacs	Van Nes and Rueb (2009)	[20]
Permeability & Integrity	More integrated areas, higher number of passersby	Decker et al. (1982)	[28]
Permeability, Intelligibility	Intelligible deformed grid streets & outward facing block	Hiller (2004)	[29]
Permeability & high pedestrian movement	Traditional through street pattern over the modern hierarchical layouts	DanzlGer (1976)	[26]
Permeability & high pederstian rate	Cul-de-sacs combined with footpath network	Montoya et al. (2016)	[30]
Intervisibility, intelligibility & high pederstian rate	Surveillable houses and accessible streets	Cozens et al. (2005)	[17]
Permeability, inter-visibility & high pederstian rate	"Cal-de-sac complex combined with footpath network"	Shu and Huang (2003)	[31]
Permeability & inter-visibility	Adjacency of streets to main routes & vital dwelling area	Change (2009)	[18]
Intelligibility	More people on the street	Shichor et al. (1979)	[27]

Following Hiller's statement with regards to the strong correlation of pedestrian movement and crime rates [26]. The pedestrian rate to a great extent was shown to be influenced by the integrity and the intelligibility of a residential neighborhood. By and Large, Integrated areas, allow easier flow and movement of people and they turn out to be less vulnerable to burglary than the segregated areas; particularly in the low and middle-income areas [21]. Contrariwise, Segregated streets, allow fewer people to enter the areas and they turn out to be more vulnerable than integrated ones. In fact, those Integrated areas that offer higher accessibility and are associated with adequate inter-visibility, can be very safe [19]. Intelligibility is another known attribute of locations with a low level of crime; having no hidden corners or dead areas. A study across six areas of a metropolitan city by [27] revealed that intelligible areas that enable more people to easily access and enter the area are less vulnerable than less intelligible areas. Indeed, houses located in more intelligible areas with good surveillability and accessibility with a greater number of people and passers-by are least at risk of burglary [17]. Moreover, the existence of passengers on accessible streets creates a vital dwelling area. The physical environment can facilitate, on street population, pedestrian rates, permeability, integrity, intelligibility, and inter-visibility of the residential neighborhood via designing an appropriate spatial setting.

Territoriality

Territoriality is "the capacity of the physical environment to create perceived zones of territorial influence" [2] with controlled access [32, 33], which is under the supervision and responsibility of specific groups of people. Territorial reinforcement uses the physical elements and certain activities to promote the sense of control and proprietorship in legitimate users of the environment and perception of risk in potential offenders [9]. It focuses on the manipulation of physical design to provide a controlling power to inhabitants for making the space less vulnerable [34]. [35] emphasized on the importance of specific architectural elements which can influence the level of social control by affecting residents' social life. The researchers who related territoriality with the low level of crime, often involved residents, dwellings, and their environment as design elements to create territorial influence in residential neighborhoods [36-39]. Concerning with dwelling design and its surroundings, 'territorial functioning' mechanism uses fences, boundaries, and other signs to mark the private areas and exclude unwanted people [35]. The demarcation of private areas as a part of the people's territory, shapes the 'defensible spaces', which are actively defended and policed by residents [2]. While the presence of "residents showing territorial concern" and "neighbors who would react"[40] will increase the perception of risk in offenders, who would want to perform any form of territorial intrusion.

In addition to the physical elements, territorial functioning has social components, such an attitude, behavior, and markers [35], which explains how people occupy, use, and manage their living spaces. A study by [21] showed that burglarized houses possessed "salient public territorial qualities" and cues of openness and unoccupied appearance. In contrast, non-burglarized houses had primary territorial qualities, such as territorial markers, communicating privacy, and inter-visibility. [41] believed that areas with higher crime level certainly require more territorial markers. Territorial access control is referred to discouraging the presence of non-legitimate users of a neighborhood by controlling the flow of vehicle and passers-by. [2] Strongly, believed in demarcating the areas, which are under the influence of inhabitants to create territorial attitudes, and added the separation of semi-public and semi-private to Jacob's basic idea of separating public zones from private zones [1]. Design should limit access paths or pathways to achieve territoriality. Empirical studies by [42] and [16] show that higher permeability of a residential neighborhood permitting more strangers increases the opportunities of property-related crimes. Hence, access control must promote territoriality to achieve safety. The design elements to achieve this sort of territoriality access control are territorial markers such as physical barriers.

Results and Analysis

Paradoxical Features of Security

Natural surveillance, access control, and territoriality complement each other, thus, designing one shouldn't decrease the strength of the other [9]. However, these strategies were taken from defensible space elements, which contain contradictions within themselves [43]. Thus, there are cases, where the design for territoriality may conflict with surveillance. For instance, physical barriers, which keep intruders out, simultaneously block visions [44]. Other physical barriers, such as walls, fences, solid gates, and landscape elements used in defining a territory reduce the inter-visibility or surveillability of one house from other neighboring houses and streets (especially on entrance and exit points). Hence, burglars are often attracted to a residence with high fences, which block neighbour's views [45]. As several contextual shreds of evidence showed that physical territorial elements, such as actual and symbolic barriers increased house vulnerability to burglary [17]. Other elements, such as 'demarcation of spaces' were supposed to evoke the sense of territoriality in an entire neighbourhood, but instead, only the private areas marked as 'defensible space' were looked over actively, whereas other areas were often ignored by the residents, so were referred as 'confused spaces'[17]. The wall marking the garden of the detached house could not add to the security against burglary, instead it could be climbed and then provide concealment [26]. For a fair comparison, the space characteristics, which have been designed according to territoriality and access control versus natural surveillance and accessibility along with their conflicting design features have been summarized in table 2.

Table 2: paradoxical design characteristics of security strategies

Territoriality and access control	Surveillance and accessorily
closed space configurations	Open space configuration
visibility to residents alone	Visibility to everyone
physical barriers	Look through barriers
Demarcation of private, neighbourhood/private and public areas	Demarcation of private and public areas
Privacy	Company
Closed environment	Open environment
Impermeable environment	Permeable environment
Close to strangers	Open to strangers
Less/Non-accessible streets	Accessible streets
Less/no pedestrian flow	More pedestrian flow
Segregated areas	Integrated areas
Cul-de-sac; dead-ends;	Through streets: Gridiron Streets
broken up streets	Connected streets
Non-permeable street pattern	Permeable street pattern

Accessibility not escapability

Offenders are very structured in selecting the location of their targets. Burglar's attention is mostly on the vulnerable entry point and available escape routes [46]. Their concern is mostly on how quickly they can both get in and get out of the target and always determine the possibilities of getting caught while in a crime scene or leaving the scene [47, 48]. Burglars make use of every available escape routes. In case of burglary, the houses must possess easy access from the street [49] or be close to major traffic routes [50]. Offenders always look for an easy escape to the nearest highway to flee quickly. Hence, street barriers, reduction of unwanted traffic, and calming the traffic with barricade/pallet increase the difficulty of offender's escape [51]. Studies show that "Through traffic, streets are far more vulnerable to residential burglaries than other types of routes, and especially T-type and dead-end"[52]. According to [53, 54], higher crime rates are seen in areas with the gridded network than areas with organic street patterns, as offenders prefer these less confusing escape routes over cul-de-sacs or complicated roads. Similarly, [42] mentioned that property crime is lower in houses located on a street with a complex network, traffic barriers, and road enclosures, which impediment access and vehicles flow. "Corner lots tend to be the most at risk for first-time burglaries" [55]. [56] also elaborated that houses at corner locations and close to the highway are more likely to be burglarized. This was further supported by [27], who showed that corner houses are more vulnerable to burglary to a significant extent. Hence, streets and the environment of residential area should control vehicle traffic and accessibility of vehicles to the houses to discourage the escapability of offenders. Study of middle-class houses by [57] asserted that the crime rate was high among the detached houses adjacent to main roads. Newman [32] in his study of Dayton's Five Oak residential community found that vehicular traffic was a major source of the neighbourhood problems. Therefore, the lack of access control and boundaries were meant to reduce the permeability of outsiders with vehicles. Hence, "the solution is to link the neighbourhood with pedestrian traffic, but disconnect from vehicular traffic" [58].

Conclusion

The main scholars in the field of CPTED believe that natural surveillance and access control must contribute to a "sense of territoriality". The empirical tests have shown strong support for natural surveillance and access control for preventing residential burglary, unlike the physical features of territoriality, such as territorial displays and extra care for the house, which increase the vulnerability of a house by assuring burglars that a target is worth and profitable. The literature indicates that social functioning of territoriality has more influence on preventing a burglary than its physical elements. Hence, the design of the residential places should include the physical aspects of natural surveillance and access control and the social aspects of territoriality to avoid the paradoxical design elements, which diminish each other's effects on burglary prevention. However, even the most exact physical design does neither guarantee the active and routine surveillance of the place and the intervention of the residents nor the control of access to the target of the offenders. Even though, the level of success of the burglary prevention strategies varies according to the contextual situations and the culture of the neighbourhoods, but the appropriate design of residential places, which consist of non-conflictory and pragmatic physical elements that facilitates good surveillance, controls the access of burglars, and motivates the sense of territoriality, enhances security to a great extent. The study recommends the examination of the paradoxical design elements of natural surveillance, access control, and territoriality in a single context to understand a more realistic level of the effects of such elements on residential burglary prevention.

Acknowledgement
This paper is part of the University Grant Scheme (RUI) (1001/PPBGN/8016013) for Quality Function Deployment for Green Building Assessment: A Sentiment Analysis with Clustering Algorithm Approach.

References

[1] Jacobs J 1961 *The death and life of great American cities*: Vintage
[2] Newman O 1972 *Defensible space*: Macmillan New York
[3] Jeffery C R 1977 *Crime prevention through environmental design*: Sage Publications
[4] Clarke R V 1980 " situational" Crime Prevention: Theory and Practice *The British Journal of Criminology* 136-47
[5] Foster S, Giles-Corti B and Knuiman M 2011 Creating safe walkable streetscapes: Does house design and upkeep discourage incivilities in suburban neighbourhoods? *Journal of Environmental Psychology* **31** 79-88
[6] Newman O 1973 *Defensible space: Crime prevention through urban design*: Collier Books New York
[7] Morgan R, Maguire M and Reiner R 2012 *The Oxford handbook of criminology*: Oxford University Press
[8] Atlas R 1991 The other side of CPTED *Security Manage*

[9] Crowe T D 2000 *Crime prevention through environmental design: Applications of architectural design and space management concepts*: Butterworth-Heinemann

[10] Newman O 1996 Creating Defensible Space. US Department of Housing and Urban Development, Office of Policy Development and Research *Institute for Community Design Analysis, Center for Urban Policy Research, Rutgers University., Washington, DC*

[11] Newman O 1980 Community of interest *Society* **18** 52-7

[12] Poyner B 1983 *Design against crime: Beyond defensible space*: Butterworths London

[13] Poyner B and Webb B 1991 *Crime free housing*: Butterworth-Architecture Oxford

[14] Hillier B and Shu S 2000 12. Crime and urban layout: the need for evidence *Secure foundations: Key issues in crime prevention, crime reduction and community safety* 224

[15] Kitchen T 2005 New urbanism and CPTED in the British planning system: Some critical reflections *Journal of Architectural and Planning Research* 342-57

[16] Painter K and Tilley N 1999 *Surveillance of public space: CCTV, street lighting and crime prevention* vol 10: Criminal Justice Press Monsey

[17] Cozens P M, Saville G and Hillier D 2005 Crime prevention through environmental design (CPTED): a review and modern bibliography *Property management* **23** 328-56

[18] Chang D 2009 Social crime or spatial crime? Exploring the effects of social, economical, and spatial factors on burglary rates *Environment and Behavior*

[19] Shu C 2009 Spatial Configuration of Residential Area and Vulnerability of Burglary: Case Studies from UK and Taiwan. In: *7th International Space Syntax Symposium, Stockholm,*

[20] Van Nes A and Rueb L 2009 Spatial behaviour in Dutch dwelling areas: How housing layouts affects the behaviour of its users. In: *Proceedings of the 7th International Space Syntax Symposium, Stockholm, Sweden, June 8-10, 2009. Eds.: Koch, D., Marcus, L., Steen, J. Trita-ARK. Forskningspublikationer, 2009: 1,*

[21] Decker D L, Shichor D and O'Brien R M 1982 *Urban structure and victimization*: Lexington Books Lexington, Mass.

[22] Hillier W and Shu S 1999 Designing for secure spaces *Planning in London* 36-8

[23] Brown B B and Altman I 1983 Territoriality, defensible space and residential burglary: An environmental analysis *Journal of environmental psychology* **3** 203-20

[24] Jacobs J 1965 *Do not segregate pedestrians and automobiles*

[25] Angel S 1968 *Discouraging crime through city planning*: Center for Planning and Development Research, Institute of Urban and Regional Development, University of California

[26] Danzinger S 1976 Explaining urban crime rates *Criminology* 14 291

[27] Shichor D, Decker D L and O'BRIEN R M 1979 POPULATION DENSITY AND CRIMINAL VICTIMIZATIONSome Unexpected Findings in Central Cities *Criminology* **17** 184-93

[28] O'Brien R M, Shichor D and Decker D L 1980 An empirical comparison of the validity of UCR and NCS crime rates *Sociological Quarterly* **21** 391-401

[29] Hillier B 2004 Can streets be made safe? *Urban Design International* **9** 31-45

[30] Montoya L, Junger M and Ongena Y 2016 The relation between residential property and its surroundings and day-and night-time residential burglary *Environment and behavior* **48** 515-49

IOP Conf. Series: Materials Science and Engineering **636** (2019) 012017 doi:10.1088/1757-899X/636/1/012017

[31] Shu S and Huang J 2003 Spatial configuration and vulnerability of residential burglary: A case study of a city in Taiwan. In: *Proceedings, 4th International Space Syntax Symposium, London*

[32] Aantjes F 2012 Residential burglaries: a comparison between self-report studies of burglars and observational data from Enschede

[33] Newman O 1975 Reactions to the "defensible space" study and some further findings *International journal of mental health* **4** 48-70

[34] Wekerle G R and Whitzman C 1995 *Safe cities: Guidelines for planning, design, and management*: Van Nostrand Reinhold Company

[35] Katyal N K 2002 Architecture as crime control *Yale Law Journal* **10** 39-139

[36] Antario C 2000 Crime Prevention Through Envionmental Design.

[37] Geason S and Wilson P R 1989 *Designing out crime: Crime prevention through environmental design*: Australian Institute of Criminology Canberra

[38] Taylor R B 1988 *Human territorial functioning: An empirical, evolutionary perspective on individual and small group territorial cognitions, behaviors, and consequences*: Cambridge University Press

[39] Taylor R B and Harrell A 1996 *Physical environment and crime*: US Department of Justice, Office of Justice Programs, National Institute of Justice Washington, DC

[40] Brower S, Dockett K and Taylor R B 1983 Residents' perceptions of territorial features and perceived local threat *Environment and Behavior* **15** 419-37

[41] Brown B B 1985 Residential territories: Cues to burglary vulnerability *Journal of Architectural Planning and Research* **2** 231-43

[42] Taylor R B and Shumaker S A 1990 Local crime as a natural hazard: Implications for understanding the relationship between disorder and fear of crime *American journal of community psychology* **18** 619-41

[43] Beavon D J, Brantingham P L and Brantingham P J 1994 The influence of street networks on the patterning of property offenses *Crime prevention studies* **2** 115-48

[44] Brown B B and Bentley D L 1993 Residential burglars judge risk: The role of territoriality *Journal of Environmental Psychology* **13** 51-61

[45] Taylor R B 2003 Crime Prevention through Environmental *Handbook of environmental psychology* 413

[46] Hirschfield A, Newton A and Rogerson M 2010 Linking Burglary and target hardening at the property level: new insights into victimization and Burglary protection *Criminal Justice Policy Review* **21** 319-37

[47] Steventon G 1996 Defensible space: a critical review of the theory and practice of a crime prevention strategy *Urban Design International* **1** 235-45

[48] Phelan G 1977 Testing "academic" notions of architectural design for burglary prevention: How burglars perceive cues of vulnerability in suburban apartment complexes. In: *annual meeting of the American Society of Criminology,*

[49] Bennett T and Wright R 1984 *Burglars on burglary: Prevention and the offender*: Gower Aldershot

[50] Ekblom P, Armitage R, Monchuk L and Castell B 2013 Crime Prevention Through Environmental Design in the United Arab Emirates: A Suitable Case for Reorientation? *Built Environment* **39** 92-113

[51] MacDonald J E and Gifford R 1989 Territorial cues and defensible space theory: The burglar's point of view *Journal of Environmental Psychology* **9** 193-205

[52] Mawby R I 1977 Defensible space: A theoretical and empirical appraisal *Urban Studies* **14** 169-79

[53] van Nes A 2005 Burglaries in the burglar's vicinity. In: *5th international space syntax symposium,* pp 479-93

[54] Taylor R B 1996 Neighborhood responses to disorder and local attachments: The systemic model of attachment, social disorganization, and neighborhood use value. In: *Sociological Forum*: Springer pp 41-74

[55] Rengert G 1980 Spatial aspects of criminal behavior *Crime: A spatial perspective* 47-57

[56] Hough M 1987 Offenders' choice of target: Findings from victim surveys *J Quant Criminol* **3** 355-69

[57] Brantingham P L and Brantingham P J 1984 *Coping with burglary*: Springer pp 77-95

[58] Atlas R 2008 21st century security and CPTED. Boca Raton, Florida: CRS Press. http://www. crcpress. com

Design Strategies in Science Centre

N Baniyamin, Z Abd. Rahim

Dept. of Applied Arts and Design, Dept. of Architecture
Kulliyyah of Architecture and Environmental Design
International Islamic University Malaysia

nurhaya@iium.edu.my

Abstract. This study presents semiotics semantics theory as a possible standard for exploring the interrelations amid learning and the design of informal learning environments. The theoretical notions of classification (Bernstein, 1996), formality (Halliday, 1996), and framing (Bernstein, 1996) were sourced to explain the approach in selected exhibition galleries in Petrosains. Studies on the relationships between signs and symbols and what they represent, especially the relations between form and their reference in the physical world or the world of ideas, have been performed. Applying the semiotics semantics theory to the science centre settings has significant ability for expanding awareness of the role that exhibition plays in the visitor experience, and apprising the design strategies. The findings suggest that the representational modes do promote an investigative learning of science in pedagogical functioning.
Keywords: informal learning setting, science centre, semiotic semantics theory, visitor experience

1. Introduction

Science centre makes science accessible to a broad range of people in engaging and enjoyable ways. Given the current climate in formal educational institutions, the need to promote and sustain informal learning environments, is even more critical. Research into relation of exhibition design and learning processes is important to understand the conditions that facilitate an enjoyable and productive learning experience (Packer, 2006). Informal learning environment offers broad, reflective, and immersive learning experiences that cannot be gained from a textbook or a classroom lesson.

2. Case Study Background

Operating since 1999, Petrosains – The Discovery Centre, is known as a high-tech educational centre dominantly exhibiting the discovery of oil and gas industry. The main objective is to enhance and create interest for the public in knowing more about petroleum science and technology. The interactive discovery centre is compacted with fun-filled activities, arranged according to the same concept as the walk-through museum which exhibits a collection of science-related elements starting from the space system to the fossils industry. The visitors are welcomed in a well-arranged and systematic route which consists of about 19 interactive showcases. For the purpose of this study, six selected exhibitions have been identified. Refer to Appendix 1.

3. Literature Review

A learning theory that combines into the motivation value is eventually the one that produces affective experiences with the construction of meaning. Affective experiences are an essential part of learning,

Content from this work may be used under the terms of the Creative Commons Attribution 3.0 licence. Any further distribution of this work must maintain attribution to the author(s) and the title of the work, journal citation and DOI.
Published under licence by IOP Publishing Ltd

reasoning, and social performance (Norman, 2004). Affect is identified as a key feature of the visitor experience (Roppola, 2012). Furthermore, this affective dimension was firmly connected to a motivation of seeking "richness of experience" comprising both entertainment and educational motivations from a museum visit (Moreno Gil & Ritchie, 2009). This suggests a possible link between affective responses and 'Learning for Fun' experiences as described by Packer (2006). Constructive affect aids vision and ingenuity, while appealing objects (which produce constructive affect in the user) are easier to use (Norman, 2004). This indicates that design as able to inspire constructive affect, and hence, better cultivate an attitude that is contributing to discovery, exploration, and learning. Further research into relation of exhibition design and learning processes is significant to value the settings that assist an enjoyable and productive learning experience (Packer, 2006). There is a need to understand the interrelations between learning and the design of learning environments. This is important because the gap between deeper interpretive contemplation of science phenomena and the experiential pleasure of science exhibits influence visitor interests and future visits.

4. Research Methodology

Enhancing visitor experiences study approach typically depend on full insights about setting and visitor behaviours. These understandings are basic to increase designers' responsiveness. Visitor behaviours were not discussed in particular, as the subjects of inquiry are process based. A purposeful sampling strategy was implemented in the selection of Petrosains, in efforts to provide an "information-rich" case from which the most could be learned (Merriam, 1998; Patton, 1990). Petrosains was chosen as the case study on the criteria of quality standard exhibition design as well as offering integration of a variety of new exhibits within its exhibition environment. Likewise, the selection was based on several specific criteria (Merriam, 1998) or attributes: size and location, quality of exhibitions, variety of exhibition elements, and the range of exhibitions present in the science centre which implements an assortment of informal learning strategies and provides multi-sensory opportunities for visitors.

5. Strategy of photography-based data collection

Data collection methods for this study included photographic analysis of the science centre's environment. The procedures begin when 20 applied arts students participated as respondents. Analysis through observation study and photography-based data collection was done during a visit from Petrosains in March 2018. Based on the limitation of time and budget factors, the analysis on degree of linearity, two-dimensional, sound, and graphic text were excluded. Only functions on three dimensional representational systems and lighting were analyzed, being totally conscious of the necessity to extend the studies to accommodate other representational modes. Photography-based data collection method is particularly strong in three aspects: capturing contextual information, reviving memory, and facilitating reflection (Rose, 2007). The photographs also refresh the respondents of exhibitions physical context, thus, prevent the loss of important information. Therefore, photographic medium is an appropriate method for this research. The data collection focused on capturing contextual data of architectural style, general interior atmospherics, space layout, and design as well as individual exhibits and displays. As photos were taken to facilitate reflection, rich contextual information allows the respondents to document what had been seen. An analytical strategy (Koulaidis et.al., 2002) based on cognitive effects of the exhibits' design in a science centre was used. The theoretical notions of classification (Bernstein, 1996), formality (Halliday, 1996), and framing (Bernstein, 1996) were sourced to explain the approach in which the exhibits support a visitor to the knowledge implication and to analyze data for identifying the multiplicity of representational modes.

6. Analytical Template

The types of concerns for the representational modes was preliminarily analyzed by organizing information into the analytical template with four components of representational system, classification, formality, and framing. The photographic data of six galleries in Petrosains were organised according to the analytical template before thematic analysis was performed to understand the representational modes in physical context of the science centre. Data from analytical template of the photographic medium were systematically coded to identify:

i. The types of epistemological relationships between knowledge categories.
ii. The degree of abstraction, elaboration, and specialization.
iii. The influence over the interaction established by the exhibits.
iv. The social interfaces in the setting of science centre.

7. Findings on Three-Dimensional Representations

The site analysis and observations of six selected galleries in Petrosains suggest a variety of common references to the every-day realm of the visitors and this is mostly made by producing examples of the way the specialized scientific knowledge attains applications that transform our everyday lives.

7.1 Classification Modulation

The content specialization promoted by the three-dimensional representations is determined by the form and the function of each representation. According to Bernstein (1996), this determines the epistemological relationship between knowledge categories of specialized 'techno-scientific knowledge' and the 'everyday knowledge'.

a. Form

Observations on six selected galleries in Petrosains suggest a balance of conventional, hybrid, and realistic representations. The three-dimensional representations promote balance of strong and weak levels of classifications in their form. For example, the model exhibits are realistic and most of them can be manipulated by the visitors. This can be seen mostly at the Oil Platform as shown in figure 1. The uniforms of the oil rig's workers were even provided. Petrosains exhibits the realistic model of F1 car for the visitors as shown in figure 2.

Figure 1

Figure 2

b. Function

The three-dimensional representations are more metaphorical as shown in Appendix 1. It signifies connotations over what they literally represent. These weak classifications highlight that there are many references to the every-day realm of the visitors. Findings indicate science centre as a ‘learning-oriented entertainment experience’ consistent with what was suggested by Moscardo (1999) where an interpretative involvement create a ‘mindful’ visitor. This mindfulness can be created by interpretive practices that encourage control, interaction, and activity, as well as introducing novelty, piquing the visitors' personal interests, and properly orienting them to the interpretive subject. The five selected galleries represent weak classifications, an enlightened and unconventional method on science learning, without the densities of the specific, and weak theoretical techno-scientific conventional representative systems. Findings indicate science centre as a ‘learning-oriented entertainment experience’ consistent with what was suggested by Bitgood (2002) where “visitor attention spans are affected by how much physical and mental effort they have to put in to stay on an exhibit”.

IOP Conf. Series: Materials Science and Engineering **636** (2019) 012018 doi:10.1088/1757-899X/636/1/012018

7.2 Framing in 3-dimensional representations

The framing (Bernstein, 1996) in 3-dimensional representations is used to measure the tendency of how the relations between the visitors and the exhibits is happening and how affective they are. The findings of strong and weak framing are reinforced by the power relationships and visitors' involvement in the three-dimensional representations.

a. Power relationships

Three-dimensional representations add to the decree of the interpersonal/affective relationships between the visitors and the exhibits. The element that contributes to the formulation of the power (hierarchical) relationships between the exhibits and the visitors is the exhibits' size and the vertical angle of view. A three-dimensional representation that is large in size, can be viewed from a high and oblique angle and which cannot be touched, tends to create a feeling of diminishment and alienation to the visitor and promotes strong framing as shown in Figure 3 and 4.

Figure 3

Figure 4

The majority of the exhibits are the human-like size exhibits, which can be viewed at an eye-level and which can be manipulated by a visitor. This tends to create a feeling of familiarity and call for involvement and promotes weak framing.

b. Visitors' involvement

The three-dimensional representations promote weak framing in its visitors' involvement. The representations are mostly where visitor can manipulate the exhibit. The analysis from all the three-dimensional representations lead to the conclusion that the visitors are allowed a great deal of independence in accessing the exhibits and are also treated as socially equal partners who are highly motivated to get involved with them. The three-dimensional representations also promote a combination of strong and weak framing in different parts of the exhibitions.

8. Findings on Lighting Representation

Lighting is an intangible medium but it has an enormous influence on both the perception of physical space and upon the emotional response of those who enter these environments. These issues are addressed with a strong sense of aesthetics, functionality, and quality of light to ensure quality visitor experience.

8.1 Formality Modulation

The findings of high and low formality (Halliday, 1996) are reinforced by degree of colour realism, colour differentiation, and degree of directionality.

a. Degree of colours realism

The low levels of formality were observed in four out of six galleries. As far as their degree of colours realism is concerned, the lighting has made more galleries categorized as realistic colours.

There are fewer galleries which are categorized as unrealistic colours as in Speed and Dark Ride galleries.

b. Colour differentiation
Whereas with respect to their colour differentiation, a balanced gallery has one or two colours as shown in Figure 5. The use of blue and yellow lighting in colours leads to high formality at Speed gallery.

c. Degree of directionality
As with respect to their degree of directionality, there are more galleries which utilized focused light beam. In conclusion, high levels of formality are characterized in the majority of exhibitions.

Figure 5

8.2 Framing Modulation
The intensity of illumination of an exhibit in relation to the corresponding intensity of the surrounding exhibits represents a significance in deciding the level of framing. The observation suggests that four out of six galleries are more intensely illuminated with respect to their surroundings and there is an implied hint to the visitor to draw his/her attention. In this way, lighting increases the visitor's control and, hence, leads to strong framing.

9. Conclusion
This research identifies the sources that generate visitor responses, known as the representational modes in the exhibitions. The reason essential to the constructing of the experience is in its matter and the sources employed by the designers. The degree of abstraction, elaboration, and specialization reinforced by the form and function of the three-dimensional representations suggest majority weak classification. The specialized character of the scientific facts can nonetheless be stated in codes of low formality. Lighting has an enormous influence on both the perception of physical space and upon the emotional response of those who enter these environments. These issues are addressed with a strong sense of aesthetics, functionality, and quality of light to ensure quality visitor experience. The low levels of formality characterize lighting in its degree of colours realism and degree of directionality. Whereas with respect to their colour differentiation, majority of the galleries have high formality. The lighting promotes strong levels of framing as majority of the galleries have intense lighting. The generated sense of independence and control over the intended knowledge can display the analytical, lively, and exploratory character of science. The study supports the potential of semiotic semantics in representational modes as a design tool for advancement in informal learning environments.

10. Research contribution
Practical insights are among these research contributions. This study offered a viewpoint for deliberating design in the informal learning setting. The concept of transformational experience (Kolb, 1984) in the design of exhibition context is acknowledged. This research analyzed the representational characteristics in science centres to facilitate understanding the relation between types of science centre offerings,

design highlights, and visitor experiences. Regarding the research method, this research uses photography-based methods in researching about science centre offerings and its design emphases. The photographic medium research method captures the structure of symbolic and experiential offers. This method further explains science centre representational expressive modes. Photographic medium captures the details visually and yields physical evidence of science centre exhibition environments. This finding confirmed that the sites attract and focus the visitors' attention effectively. It is demonstrated that science centre exhibition environment is designed for affective experiences. Science centre creates positive attitudes and lets visitors learn for fun.

Acknowledgement
The authors wish to thank the International Islamic University Malaysia and the Ministry of Higher Education for funding this project in RIGS17-086-0661.

References

[1] Bernstein, B. (1996). *Pedagogy, Symbolic Control and Identity: Theory, research, critique*. London: Taylor and Francis.
[2] Bitgood, S. (2002). Environmental psychology in museums, zoos, and other exhibition centers. In R. B. Bechtel & A. Churchman (Eds.), *Handbook of environmental psychology* (461–480). New York: John Wiley & Sons.
[3] Halliday, M.A.K., (1996). On the Language of Physical Science. In M.A.K. Halliday and J.R.Martin (Eds.), *Writing Science: Literacy and Discursive Power,* (54-68). London: The Falmer Press.
[4] Kolb, D. A. (1984). *Experiential learning: Experiences as the source of learning and development.* Englewood Cliffs, NJ: Prentice Hall.
[5] Koulaidis, V., Dimopoulos, K., & Matiatos, S. (2002). Science and technology centers as 'texts'. *International Journal of Learning,* ***9***, 717-735.
[6] Merriam, S. B. (1998). *Qualitative Research and Case Study Applications in Education*. San Francisco: Jossey-Bass.
[7] Moreno Gil, S., & Ritchie, J. R. B. (2009). Understanding the museum image formation process: A comparison of residents and tourists. *Journal of Travel Research*, ***47*****(4)**, 480–493.
[8] Moscardo, G. (1999). *Making Visitors Mindful: Principles for Creating Quality Sustainable Visitor Experiences Through Effective Communication.* Champaign, IL: Sagamore.
[9] Norman, D. A. (2004). *Emotional Design: Why We Love (Or Hate) Everyday Things*. New York: Basic Books.
[10] Packer, J. (2006). Learning for fun: The unique contribution of educational leisure experiences. *Curator: The Museum Journal*, ***49*****(3),** 329–344.
[11] Patton, M.Q. (1990). *Qualitative evaluation methods* (2nd ed.). Thousand Oaks, CA: Sage Publications.
[12] Roppola, T. (2012). *Designing for the Museum Visitor Experience.* London: Routledge.
[13] Rose, G. (2007). *Visual Methodologies: An Introduction to the Interpretation of Visual Materials.* London; Thousand Oaks, CA: Sage Publications.

IOP Conf. Series: Materials Science and Engineering **636** (2019) 012018 doi:10.1088/1757-899X/636/1/012018

Appendix 1

Analysis on Three dimensional-representations and Lighting

Science Centre	**Three dimensional-representations**								**Lighting**							
Petrosains	Classifications (strong/weak)				Framing (strong/weak)				Formality (high/low)						Framing (strong /weak)	
	Form		Function		Power relationships		Visitors' involvement		Degree of colours realism		Colour differentiation		Degree of directionality		Intensity of lighting with respect to the surroundings	
	Conventional representation or hybrid	Realistic representation	Classificational, Analytical, Narrative	Metaphorical	Viewed from below/ large size	Viewed from above/ small size	Visitor can only see the exhibit	Visitor can manipulate the exhibit	Unrealistic colours	Realistic colours	One or two colours	Three or more colours	Focused light beams	Diffuse lighting	Intense lighting	Uniform lighting everywhere
Dark Ride	*	*		*	*		*		*		*		*		*	
The Known Universe	*	*		*		*		*		*	*		*		*	
Sparkz	*			*		*		*		*		*	*		*	
Oil Platform		*		*	*			*		*		*		*		*
Speed	*		*			*	*	*	*		*		*		*	
Molecule Cafe		*		*	*		*			*		*		*		*
TOTAL	4	4	1	5	3	3	3	4	2	4	3	3	4	2	4	2

IOP Conf. Series: Materials Science and Engineering **636** (2019) 012019 doi:10.1088/1757-899X/636/1/012019

COMMUNITY RAINWATER HARVESTING FINANCIAL PAYBACK ANALYSES - CASE STUDY IN MALAYSIA

Yoke Bee Woon[1], Lloyd Ling[1*], Wei Lun Tan[2], Ming Fai Chow[3]

[1]Centre of Disaster Risk Reduction (CDRR), Lee Kong Chian Faculty of Engineering & Science, Civil Engineering Department of Universiti Tunku Abdul Rahman, Jalan Sungai Long, Bandar Sungai Long, 43000 Cheras, Malaysia.

[2]Centre for Mathematical Sciences(CMS), Lee Kong Chian Faculty of Engineering & Science, Department of Mathematical and Actuarial Sciences of Universiti Tunku Abdul Rahman, Jalan Sungai Long, Bandar Sungai Long, 43000 Cheras, Malaysia.

[3]Institute of Energy Infrastructure, Universiti Tenaga Nasional, 43000 Kajang, Selangor, Malaysia.

*Corresponding Author: *linglloyd@utar.edu.my*

Abstract. Malaysian water demand is increasing at an alarming rate reaching 27 to 38% higher than the World Health Organisation recommended consumption limit of 165 liters per capita per day. Therefore, the Malaysian water shortage crisis is quite possible in future due to this water demand uptrend. The average annual rainfall of Malaysia is 2,400 mm but large portion of this fresh water resource becomes runoff and lost through our catchments. Urban flash flood is also becoming more frequent due to fast pace of urban development and anthropogenic induced runoff. Malaysia has experienced drought and flooding in different areas and therefore, it is crucial to study the feasibility of alternate water resources in Malaysia to manage and maintain the sustainability of urban township. This study reviewed a past rain water harvesting system (RWHS) case and conducted the financial payback analyses on its proposed system. If there were 177 rain days per year with at least 52 mm of rainfall event depth, the payback period of the proposed RWHS would be 5.8 years when the discount rate (i%) = 2% and 8.2 years if i% = 10%. The payback period became longer when the annual rain days dropped below 106 and 89 rain days per year. If the proposed RWHS only serve the community under this study, it will take 12 days to consume 800 m^3 stored water, while any rainfall of consecutive days will not be harvested as the underground storage tank is in full capacity. The proposed RWHS must be filled up at least 38 times per year in order to break even with the proposed annual maintenance cost but will never be able to achieve any payback from its initial investment. Rain water harvesting and full utilisation is the only way to achieve high water cost savings and shorter payback period, and maximise urban excess runoff reduction.

1. Introduction

Content from this work may be used under the terms of the Creative Commons Attribution 3.0 licence. Any further distribution of this work must maintain attribution to the author(s) and the title of the work, journal citation and DOI.
Published under licence by IOP Publishing Ltd

The increasing world population is causing the steep rise in water demand globally, while water security of our future is becoming less promising [1 2 3]. Malaysian water demand increased from 10.4 billion m^3/year in 1998 to 12.1 billion m^3/year in 2010. By 2050, the projected demand will reach 17.7 billion m^3/year [4]. On average, Malaysian water consumption ranges from 209 to 228 liters per capita per day (lcd) [5] which is about 27-38% higher than the recommended consumption limit of 165 lcd [6] set by the World Health Organisation (WHO). Therefore, the Malaysian water shortage crisis is quite possible in future due to this water demand uptrend. The average annual rainfall of Malaysia is 2,400 mm [7] but large portion of this fresh water resource becomes runoff and lost through our catchments. Urban flash flood is also becoming more frequent due to fast pace of urban development and anthropogenic induced runoff. Malaysians have experienced drought and flooding in different areas and therefore, it is crucial to study the feasibility of alternate water resources in Malaysia to manage and maintain the sustainability of urban township.

Rain water harvesting system (RWHS) is one of the alternatives to collect and utilize natural fresh water from Mother Nature, while the implementation of large scale RWHS is able to reduce the occurrence of urban flash flood up to 10% as reported by South Korean research [8] and improve the water quality of water bodies [9]. However, the general concern of adopting RWHS is the initial high upfront cost with the payback uncertainty due to the unforeseen future rainfall amount and pattern. It is also not appealing to consider RWHS as Malaysian water tariff is quite low compared to our neighbor countries. RWHS is only economical when it serves as a mechanism in water supply and flood control at the same time [10]. When implemented through proper planning, large scale RWHS or network functions like water storage infrastructure serving the water demand of urban communities.

2. Study Site and Methodology

This study reviewed past RWHS study conducted by NAHRIM on part of the community within Taman Wangsa Melawati (as shown with dashed line in Figure 1) in the Selangor state of Malaysia. 242 double storey terrace units, 10 shops, 1 mosque, and 1 kindergarten are in this community with total catchment area of 15,540 m^2. Daily water demand from the past study estimated 240 liters for each terrace unit, 375 liters for each shop, 1,100 liters for the mosque, and 2,450 liters for the kindergarten. Total daily water demand of this community was estimated to be 65.4 m^3.

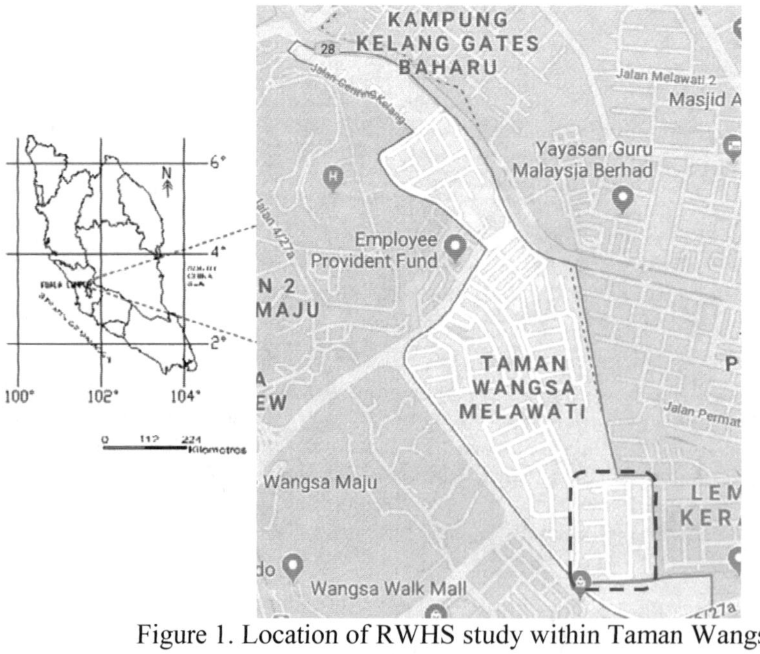

Figure 1. Location of RWHS study within Taman Wangsa Melawati

This community was chosen for RWHS study as it has a public garden in the middle for a potential installation of a 800 m^3 underground storage tank with cost estimation as shown in Table 1. The layout of the proposed RWHS is shown in Figure 2.

Table 1. Overall Cost Estimation of the RWHS

No.	Items	Estimated Cost (RM)
1	Plumbing and Piping	770,750
2	Pumps and Fittings	50,000
3	Underground Storage Tank	208,000
4	Elevated Distribution Tank	176,000
	Overall Cost:	**1,204,750**
	Projected Annual Maintenance (5% of overall cost):	60,237.50

According to one of the studies, Selangor state water tariff for domestic use is RM 2/m^3 and RM 2.28/m^3 for non-domestic accounts, while on average, there were 177 rain days/year in Kuala Lumpur area between 1953 and 2008 [5]. Therefore, this study adopted the past rain days/year statistics to assume that there will also be at least 177 rain days/year in this community in future. From the total catchment area of this community, rainfall event depth must be greater than 52 mm in order to be able to harvest 800 m^3 of rainwater to fill up the proposed underground storage tank, which will be sufficient to cater to the consumption of this community for 12 days. In the event that the proposed underground storage tank cannot be filled up, the anticipated water savings from this RWHS will be less promising with longer payback period.

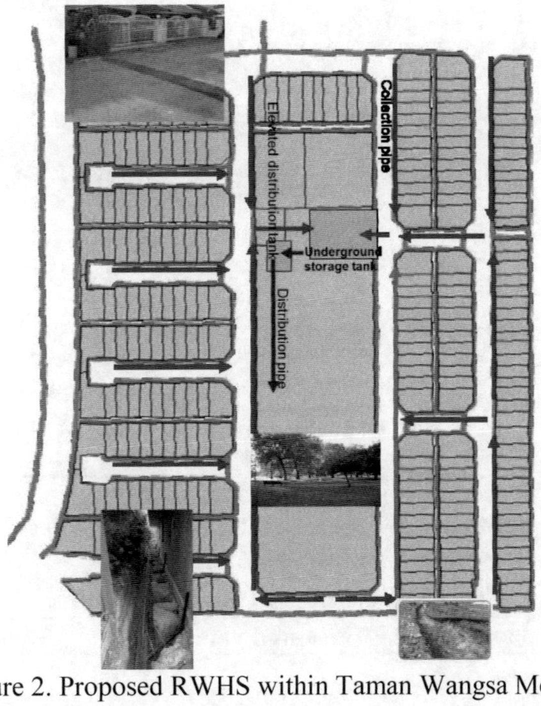

Figure 2. Proposed RWHS within Taman Wangsa Melawati

This study used the minimum water tariff cost of RM 2/m^3 to quantify the projected water saving cost in future with rainfall event of 52 mm or more under three different scenarios. Nearby communities will tap into this storage tank and consume 800 m^3 stored water within 2 days. **Scenario one:** There will be 177 rain days/year with 177 rainfall events of 52 mm rainfall depths or more in future. **Scenario two:** There will be 106 rain days/year (to represent the 40% below average condition) with 106 rainfall events of 52 mm rainfall event depths or more in future and **Scenario three:** There will only be 89 rain days/year with 89 rainfall events of 52 mm rainfall event depth or more in future to represent the 50% below average raining days/year scenario for this community. Compound interest rate financial analyses was conducted under multiple discount rates (i% from 2% to 10%) in order to calculate the payback period (years) for this RWHS proposal under aforementioned three different scenarios from rain water harvesting only.

3. Results and Discussion

The RWHS harvested a total of 141,600 m^3/year of rain water under the first scenario through 177 rain days. Water tariff cost of RM 2/m^3 produced RM 283,200/year savings for the community (Refer to Table 2 for the overview of annual savings of the RWHS under three different scenarios). After the discount of the RM 60,237.50 annual maintenance cost for the RWHS, the net annual cash flow is RM 222,962.50. The payback period of the RM 1.2 million RWHS investment under different discount rates showed 5.8 years when i% = 2% and 8.2 years if i% = 10%. However, the payback period of the proposed RWHS became longer under the other two scenarios when there were less rain days/year with at least 52 mm rainfall event depths. Scenario two showed that the RWHS would take 12.6 years to breakeven if i% = 2% and 55.5 years if i% = 9%, while the payback period under scenario three showed that the RWHS would take 17.5 years to breakeven if i% = 2% and 36.3 years to breakeven even if i% is only 6%.

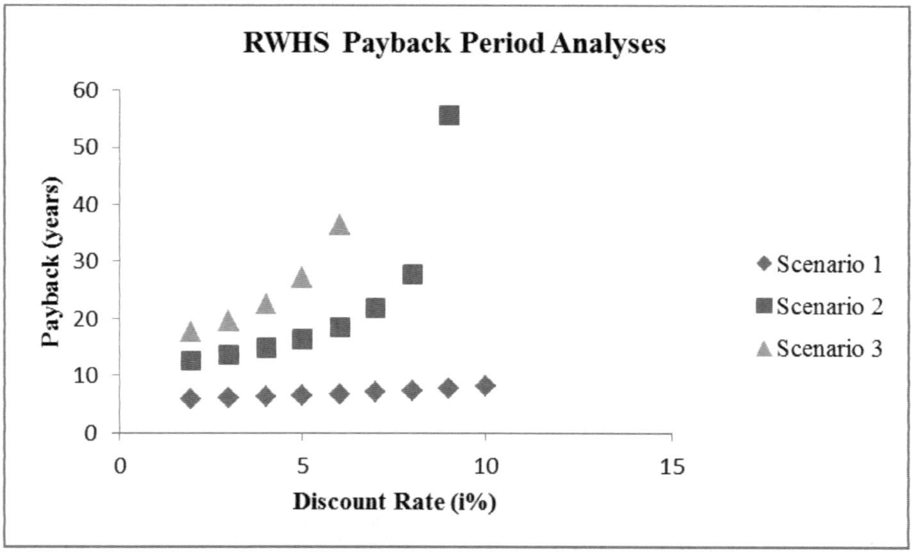

Figure 3. Payback period of the proposed RWHS within Taman Wangsa Melawati

Table 2. Overview of the RWHS under 3 scenarios

Scenario	Rain days/year*	Harvested volume (m^3)	Annual Savings (RM)
1	177	141,600	283,200
2	106	84,800	169,600
3	89	71,200	142,400

*Minimum rainfall depth is 52 mm

4. Conclusions

The community under this study is densely populated as many other urban areas in Malaysia. If there were 177 rain days with at least 52 mm of rainfall event depth, the payback period of the proposed RWHS would be very promising provided that the discount rate of the initial RM 1.2 million investments can be secured below 10%. There will be no payback for scenario 2 if the discount rate is more than 9% and no payback for scenario 3 if the discount rate is more than 6%.

The payback period also became longer when the annual rain days dropped below average. However, it is noteworthy to mention that although the total catchment area of this community under this study was only 15,540 m^2 the RWHS was able to harvest nearly 71,000 to 85,000 m3 of rain water annually under the scenarios two and three to serve the water demand of the community and its nearby communities under this study.

If the proposed RWHS only serve the community under this study, it will take 12 days to consume 800 m3 stored water while any rainfall of consecutive days will not be harvested as the underground storage tank is in full capacity. The proposed RWHS must be filled up at least 38 times per year in order to break even with the proposed annual maintenance cost but will never be able to achieve any payback from its RM 1.2 million investment. Rain water harvesting and full utilisation is the only way to achieve high water cost savings and shorter payback period, and maximise urban excess runoff reduction.

References

[1] Vörösmarty, C.J.; Green, P.; Salisbury, J.; Lammers, R.B. Global water resources: Vulnerability from climate change and population growth. *Science* **2000**, *289*, 284–288.

[2] Ercin, A.E.; Hoekstra, A.Y. Water footprint scenarios for 2050: A global analysis. *Environ. Int*. 2014, 64, 71-82.

[3] De Fraiture, C.; Wichelns, D. Satisfying future water demands for agriculture. *Agric. Water Manag*. **2010**, *97*, 502–511.

[4] New Development and Challenges in Malaysian Drinking Water Supply. Available online: http://slideplayer. com/slide/5175607/ (accessed on 25 March 2018).

[5] Lani, N.M., Yusop, Z. and Syafiuddin, A. A Review of Rainwater Harvesting in Malaysia: Prospects and Challenges. *Water*. **2018**, 10, 506.

[6] Water Efficiency. Available online: http://www.awer.org.my (accessed on 20 November 2017).

[7] Che-Ani, A.; Shaari, N.; Sairi, A.; Zain, M.; Tahir, M. Rainwater harvesting as an alternative water supply in the future. *Eur. J. Sci. Res*. **2009**, *34*, 132–140.

[8] Kim, K.; Yoo, C. Hydrological modeling and evaluation of rainwater harvesting facilities: Case study on several rainwater harvesting facilities in korea. *J. Hydrol. Eng*. **2009**, *14*, 545–561.

[9] Rahman, A.; Keane, J.; Imteaz, M.A. Rainwater harvesting in Greater Sydney: Water savings, reliability and economic benefits. *Resour. Conserv. Recycl*. **2012**, *61*, 16–21.

[10] Fewkes, A. A review of rainwater harvesting in the UK. *Struct. Surv*. **2012**, *30*, 174–194.

IOP Conf. Series: Materials Science and Engineering **636** (2019) 012020 doi:10.1088/1757-899X/636/1/012020

Employee Turnover of Quantity Surveying Firms in Malaysia

Sheikh Ilmi, Habizah[1], He Xue Qi[1] and Meor Gheda, Mariatul Liza[2]

[1]School of Architecture, Building and Design, Taylor's University, Malaysia
[2]Cluster of Applied Science, Open University, Malaysia

Email: Habizah.ilmi@taylors.edu.my

Abstract. Employees are important stakeholders in any firm. There is a trend of a high employee turnover rate in many firms in the construction industry. However, there is lack of emphasis on the reasons of employee turnover especially in the quantity surveying (QS) firms in Malaysia. The purpose of this paper is to examine the factors that contribute to turnover and its impact to the quantity surveying consultant firms. The data was collected by distributing survey questionnaires to employees of QS firms to obtain their views on the subject matter. From the analysis, the factors that caused employee turnover were established. Three quarters of respondents replied that their firms are experiencing staff turnover and the most significant impact to the firms is that they are losing their experienced quantity surveyors. In addition, QS staffs are dissatisfied to their current salary and this factor causes them to seek for a higher pay in other firms or places.

1. Introduction

In Malaysia, the Ministry of Human Resources in their Labour and Human Resources Statistics (2012) showed that the retrenchment of labour in construction industry increased significantly from 169 in year 2011 to 1002 in year 2012. A survey done by Aon Hewitt in 2015, Malaysia recorded the second highest involuntary turnover rate at 6.0% and the third highest voluntary turnover rate at 9.5% in South East Asia. There was also a downward trend of average salary increase from 6.4% to 6.2% from 2015 to 2016 (projection) in the construction or engineering industry (HR in Asia, 2015). In 2017, the average turnover rate was at 20% which was considered high and the involuntary turnover still represented 6% in overall (Institute of Labour Market Information and Analysis (ILMIA), 2018). Furthermore, about 8% out of total 645 employees in the land and quantity surveying professional services industry was the outsourced professionals (ILMIA, 2018).

According to Dias (2012), there are two types of turnovers; voluntary turnover and involuntary turnover. Voluntary turnover is initiated by the employee for many different reasons and it can be predicted and addressed in human resources. As for involuntary turnover, it is where the employees have no choice in their termination such as an employer-initiated due to non-performance of an employee. According to the survey by the Randstard (2011) from year 2011 to 2012, it reported that 55% of construction players have been working with their current employer for only less than two years, yet 85% of them plan to leave their job in the next 12 months. Many consultants will choose to change their working environment after having tried working in the construction industry due to the workload and longer working hours (Wong et.al, 2010). Kokkaew (2012) described that due to the diverse, complex and risky nature of the construction industry, there are less young people that are attracted to enter the construction industry, thus, contributing to the lack of skilled workers problem.

Content from this work may be used under the terms of the Creative Commons Attribution 3.0 licence. Any further distribution of this work must maintain attribution to the author(s) and the title of the work, journal citation and DOI.
Published under licence by IOP Publishing Ltd

Furthermore, the shortage of professionals in construction industry will also lead to employee turnover in construction industry (Hee and Ling, 2011). Phillips & Connell (2003) mentioned that "in almost all segments of the economy, economic expansions translate directly into new jobs, which in turn create new opportunities for employees to leave current employment". Therefore, the economic growth also partially contributes to employee turnover.

Construction team players opt to leave their current job due to several reasons, such as no increment of salary, being dissatisfied with their working environment, corporate politics issues, etc. (Bilau et al., 2015). According to the Staff (2014), it is insisted that poor relationship with the boss is the primary key for the staff to quit their firm in the construction industry. Beswick et al. (2007) found that 87 out of 1732 people involved in UK's construction industry experience stress, depression, or anxiety, which they believed was caused by their job. Majority of them disclosed that one of the most stressful aspects of work is having too much work to do in one time due to tight deadlines and powerless authority to choose or change the order of tasks and method of working, causing them to work longer hours. Furthermore, it is common for an individual to handle two to three projects at a time in the construction industry. Due to this heavy workload, some of them may opt to convert to another firm with a more flexible offer (Leung et al., 2005). In addition, the job scopes of construction players are dangerous or risky as they need to juggle between the projects – going to construction sites, supervising work, be present for meetings, and valuing work done (Beswick et al,. 2007). Danso (2012) discovered that individuals working in the construction sector in Ghana are significantly dissatisfied with their work benefits such as payment of allowances or bonuses besides salary and additional annual leaves.

Staff turnover in the construction industry will delay the period of completion, affect project quality, incur client dissatisfaction, and arise profit issue (Cooper, 2004). To reduce the intention of turnover, employers or the human resource managers have to find out the solutions so that the turnover rate in a company can be reduced (Dainty and Loosemore, 2012). Staw, Bell and Clausen (1986) insisted that the manager and the organization have to focus on employee's satisfiers and dissatisfiers in order to help to retain the key people in an organization (as cited in Arokiasamy, 2013).

Despite of many literature from previous studies, the findings replete in general about employee turnover. However, there is lack of emphasis on the reasons of employee turnover in the quantity surveying (QS) firms in Malaysia. Therefore, this study will fill a gap by identifying the reasons of employee turnover in the QS firms. Every organization wishes to have high productivity, few turnovers, and more profit. Managing turnover successfully is a must to achieve these goals. Thus, it is very important to find out what the reasons are that contribute to employee turnover in consultant firms so that managers or the organization can establish solutions to overcome this issue.

2. Research Method

This study is based on the quantitative approach where the questionnaire survey is used to measure objectively the causes and impact of employee turnover to QS firms. The questions were designed based on the information gathered from the literature review. The design of the questionnaire is set to get the answers from the registered quantity surveyors who are working in the QS consultant firms. They are the important sources as they can provide the information and are related to the real issues of this research. The population is obtained from the Board of Quantity Surveyors Malaysia (BQSM) website, which is the official website to get information about the QS in Malaysia. There are 364 QS consultant firms in Malaysia. However, the BQSM website does not classify whether the registered quantity surveyor is working as a consultant quantity surveyor or a contractor's quantity surveyor. Hence, the scope of study area has been reduced to focus on only the state of Selangor, and the list of QS consultant firms which was amounted to 120 was retrieved from the BQSM website. The researchers have called the firms to enquire the number of registered quantity surveyors working in

IOP Conf. Series: Materials Science and Engineering **636** (2019) 012020 doi:10.1088/1757-899X/636/1/012020

their respective firms. It has been established that there are 254 consultant registered quantity surveyors working in the consultant firms in Selangor.

3. Results

Out of the total 254 questionnaires sent out, the response rate received was only 55 or 21.65%. Based on the data gathered, 31% of the respondents work in the industry for more than 21 years and above, 22% of the respondents have worked for 11-15 years, and followed by 19% of respondents who have worked for 16-20 years. Whereas 16% of the respondents have worked less than 5 years and the remaining 12% was held by the respondents who have worked for 6-10 years. Therefore, we can conclude that 70% of the respondents are working more than 11 years, making them more experienced than others as they have already experienced and witnessed how the employee turnover affect their working life. Hence, their answers are more relevant and useful for this research.

78.13% of the respondents claim that their company is facing the employee turnover and 44% of respondents indicated that there are 3-4 QS staff who will leave their company in a year. 41% of the respondents claimed that there are 1-2 QS staff who will quit the firm and 9% of respondents claimed that 5-6 QS staff will leave their company in a year. There is also 6% of the respondents who agreed that 7 and more QS staffs are leaving in a year. It can be deduced that most of the QS firms are experiencing a scenario that a few staff are leaving their firm every year. The data also shows that 56% of the respondents indicated that they would stay in a company for average 1-5 years. 22% of the respondents reported that they would stay in a company for around 6-10 years. There are 16% of the respondents who insisted that they will stay in a company for more than 10 years and above. However, there are 6% of the respondents replied that they will not stay in a company for more than a year. We can conclude that in few year times, the quantity surveyors are more inclined to change their working environment and choose to work at other company.

Table 1 shows the respondents' level of satisfaction in their firm. The respondents show a good satisfaction with their working environment and job. Employees that are working under their preferred working environment will increase their work commitment and lessen complaints and absenteeism. The data in Table 1 shows a significant attitude of the employees against the organizational commitment and the boss or the team leader. The result shows that this segment of employees is at the borderline of their satisfaction and they can leave the job with any minor encouragement received from the other employers. As expected, most of the respondents are not satisfied with their salary and employee benefits, thus, making them rank lower as compared to other variables.

Table 1: QS's level of satisfaction

Item	Level of QS's Satisfaction	Mean	Rank
1	Working Environment	3.78	1
2	Job	3.72	2
3	Organizational Commitment	3.41	3
4	Boss or Team Leader	3.41	3
5	Salary	3.28	5
6	Employee Benefit	3.25	6

Table 2 shows the significant factors which contribute to the employees' decision to quit their current company or construction industry. It illustrates that higher pay from other companies is the most significant factor that the quantity surveyors consider in deciding whether they want to quit their current company. Due to the economic factor and money value, it is very common for workers to switch to a higher paid job since the work scope is the same. The salary and level of remuneration are the most important factors to motivate the workers to stay in their company. Alternative job opportunity means the quantity surveyors do not only work in consultant firms, they also have other options to choose either to work with developers or contractors or become academicians. Therefore, there are many job opportunities they can choose if they get bored with their current job and this will benefit most to the experienced quantity surveyors who will easily find a new job. Thus, alternative job opportunity is ranked as the second important factor beside the salary that they will get paid.

Table 2: Causes of employee turnover of QS consultant firms

Item	Reasons of Employee Turnover	Mean	Rank
1	Higher pay from other companies	3.83	1
2	Alternative job opportunity	3.75	2
3	Current working environment	3.05	3
4	Heavy workload and stressful life	3.03	4
5	Better employee benefits from other companies	2.85	5
6	New company is able to provide you challenges so you can achieve more goals in the working life	2.83	6
7	Your spouse is transferred to another state	1.70	7
8	Chances to practice religious beliefs	1.46	8

Even though the current working environment is not ranked among the most significant causes, the quantity surveyors will start to think to find a preferred working environment if they work under unpleasant working environment in addition to staying at office for 8-9 hours during working days. Hence, bad working environment will affect their mental and physical conditions. Besides salary, employee benefits are perceived as a reward to comfort the employees. The workers are most likely to stay in the company if they find out that the company could provide better employee benefits compared to other companies. In the construction industry, the quantity surveyors have a considerable amount of workload and stress. Hence, if the company can provide staff benefits, such as insurance, ex gratia payment or yearly bonus and company trips, this will be the pull factor to retain their staff. Some people will seek to obtain their working life goal but this does not apply to all. Some employees want more challenging works and make a significant contribution to the firm. If they find that they are not fond of their current job and repeating their daily work life at the same pace, they are more likely to look for another job. However, some employees will choose to quit if they are facing new challenges and eventually not coping well with them. For example, there are senior quantity surveyors who do not know how to utilize the measurement software and this will ultimately cause them to voluntarily change their job scope, in which software application is not a requirement in their duties. The factors, such as dedicating themselves in their marital status, having the spouse transferred to other places, and chances to practice religious beliefs are not the significant causes of quitting the current job. If the staff cannot cope with family and job commitment at the same time, changing job is likely to happen.

It was found out that the loss of experienced QS staff has the most significant impact to the company (refer to Table 3). This is due to the fact that experienced quantity surveyors are very hard to find in the industry and the company spends longer time to train a freshly graduated QS to become an experienced quantity surveyor. Moreover, the experienced quantity surveyors are able to handle few projects at the same time. If they quit the company, their on hand works will be transferred to the staff that still remain in the company. Hence, the second highest impact is workload increase to the existing staff. The boss or the team leader will assign another quantity surveyor or new staff to take over the projects. Overwork will bring serious impacts such as error in cost estimation which will result in reduction of the efficiency and productivity of the firm. This is due to the person who take over the work is still new to the project. Hence, he/she may make mistakes which will affect the productivity of the works. The employee turnover incurs some costs in the replacement, recruitment, and selection of staff. If there is high employee turnover, the company will need to make advertisements in order to recruit new employees. Besides, the employees who remain in the company will claim for overtime pay in order to finish the additional workload. Hence, most respondents highly agreed that employee turnover incurred additional cost.

Table 3: Impact of employee turnover to the QS firms

Rank	Impacts	Average Index
1	Loss of experienced QS staff	4.28
2	Increasing workload of other staff who stayed in the firm	4.16
3	The efficiency and productivity of the firm is reduced	4.03
4	Incurred cost in the replacement, recruitment, and selection of staff	3.91

4. Conclusion

In a nutshell, a company which is experiencing employee turnover will face the most serious effect, i.e. they are losing their experienced employee. An experienced employee needs time and cost to train. When a senior quantity surveyor resigns, the employer will have to incur cost in reemployment, assign a new or inexperienced staff to take over, and acknowledge to other parties of the project especially the client and contractor. If the client always receives the notice of frequent exchange of QS consultant, this will cause them lose their confidence towards the firm. Hence, the significant impact of losing an experienced quantity surveyor can be a serious damage to the QS consultant firm and bring other impacts as well.

In order to reduce the employee turnover in the QS firms, it is suggested that the company should provide bonus and motivating rewards to retain the staff. Providing employee benefits, such as vacation, medical insurance, and retirement plans are necessary to cater the needs of the workers. Monetary benefits such as an increase of salary can retain the staff as they are satisfied with their wages, who are likely to remain with the organizations. Improving working environment will help the staffs to maintain their commitment and sense of belonging to their company. Providing training to the QS staff also can reduce the turnover rate as well. Implementation of Building Information Modelling (BIM) software can reduce the workload of QS staff and improve the efficiency of works. Hence, if the staff have the chances to learn new skills, they would like to stay in the current company and contribute their expertise. Lastly, allowing the staff's involvement in the decision making can effectively reduce the turnover rate.

References

[1] Aon Hewitt (2017). Technology Companies in Asia-Pacific Ready to Boost 2018 Salary Budgets to Address Rising Employee Turnover And Strong Demand for Talent. 1 November 2017. Available at https://apac.aonhewitt.com/document-files/media/november-2017/radford-2017-press-release-apac.pdf.

[2] Arokiasamy, A. R. A. (2013). A qualitative study on causes and effects of employee turnover in the private sector in Malaysia. Middle East Journal of Scientific Research. 16(11). 1532–1541. Available at: http://www.scopus.com/inward/record.url?eid=2-s2.0-84887770208&partnerID=tZOtx3y1.

[3] Beswick, J., Rogers, K., Corbett. E., Sarah, B. and Jackson, K. (2007). An analysis of the prevalence and distribution of stress in the construction industry, Health and Safety Executive.

[4] Bilau, A. A., Ajagbe, M. A., Sholanke, A. B. and Sani, T. A. (2015). Impact of employee turnover in small and medium construction firms : A Literature Review. International Journal of Engineering Research & Technology (IJERT). 4(02), 976–984.

[5] Cooper, D. (2004). Improving People Performance in Construction. Gower Publishing Limited.

[6] Dainty, A. and Loosemore, M. (2012). Human Resource Management in Construction: Critical Perspectives (2nd ed). New York: Routledge.

[7] Danso, H. (2012). Construction Workers' Satisfaction with Work Provision Requirement Dimensions in Ghana's Construction Industry. International Journal of Engineering and Technology. 2(9), 1613–1619.

[8] Dias, L. P. (2012). Beginning Management of Human Resources, Washington: Flatworld Knowledge.

[9] Hee, C. H. S. and Ling, F. Y. Y. (2011). Strategies for reducing employee turnover and increasing retention rates of quantity surveyors. Construction Management and Economics. 29(10), 1059–1072.

[10] HR in Asia (2015), Key Insights to Attract, Reward and Retain Talent in Malaysia: Aon Hewitt's Views. Online. Available at: http://www.hrinasia.com/recruitment/key-insights-to-attract-reward-and-retain-talent-in-malaysia-aon-hewitts-views/, 3 December 2015.

[11] Institute of Labour Market Information and Analysis (ILMIA), Ministry of Human Resources (2018), *Jobs, Salaries And Vacancies, 2017.* Online. Available at https://www.ilmia.gov.my/index.php/en/research-publication/publication.

[12] Institute of Labour Market Information and Analysis (ILMIA), Ministry of Human Resources (2018), *Exploring The Labour Market: A Snapshot of ILMIA's 2018 Initiatives*. Online. https://www.ilmia.gov.my/index.php/en/research-publication/publication.

[13] Kokkaew, N. (2012). Current Practices of Human Resource Management (HRM) in Thai Construction Industry : A Risk and Opportunity Perspective. Review of Integrative Business & Economics. 1(1), 1–14.

[14] Leung, M. -Y., Olololaiye, P., Chong, A., Lam, C. C. Y. (2005). Impacts of stress on estimation performance in Hong Kong. Construction Management and Economics. 23(9), 891–903.

[15] Ministry of Human Resources (2012). Labour and Human Resources Statistics (2012).

[16] Phillips, J.J. and Connell, A.O., 2003. Managing Employee Retention : A Strategic Accountability Approach. Burlington: Butterworth-Heinemann.

[17] Staff, E. (2014). Bad Boss Relationship Spurs Most Turnover. ENR: Engineering News Record 1.

[18] Randstard (2011). Randstad World of Work Report.

[19] Staw, B. M., Bell, N. E. and Clausen, J. A (1986). The Dispositional Approach To Job Attitudes : A Lifetime Longi- tudinal Test. Administrtive Scienece Quartely. 31(1),56–77.

[20] Wong, J., Teo, M. and Cheung, F. (2010). Cultural Determinants of Stress in the Construction Industry. In Proceedings of 2010 International Conference on Construction & Real Estate Management. Brisbane, Queensland.

IOP Conf. Series: Materials Science and Engineering **636** (2019) 012021 doi:10.1088/1757-899X/636/1/012021

Degree Level of Publicness Through Meaning of Public Sphere In Bandung City, West Java, Indonesia

Case Study: Alun-alun Square, Bandung City, West Java, Indonesia

Doddy Friestya Asharsinyo[1)]
Lecturer of the Faculty of Creative Industry, Universitas Telkom, Bandung, Indonesia.
Email: doddyfriestya@telkomuniversity.ac.id
Ully Irma Maulina Hanafiah[2)]
Lecturer of the Faculty of Creative Industry, Universitas Telkom, Bandung, Indonesia.
Email: ullyrmaulinafia@telkomuniversity.ac.id
Muhizam Mustafa[3)]
Lecturer of School Of Arts, University Sains Malaysia, Malaysia
Email: mmuhizam@usm.my
Mohd Hafizal Mohd Isa[4)]
Lecturer of School Of Arts, University Sains Malaysia, Malaysia
Email: hafizal@usm.my

Abstract. Bandung city public area at this time is experiencing a rapid growth due to economic growth and the influence of globalization. The development of public spaces that tend to be limited to the people who live in them, makes it difficult for the public to enter the area of its public sphere. Public sphere are formed based on the economic, political, cultural, developmental, and changes that occur in the area of today's public sphere, making it limited and inaccessible to the public at large. This is caused by the hierarchy of public space formed by the changes of the functions, forms, and meanings. This study focused on the relationship between form, function, and meaning of public sphere to the degree level of public spaces of the city of Bandung.
The purpose of this study is to reveal all the relationships that exist between the concepts of public space and its influence on the degree level of publicness in the space of the city of Bandung in the context of its change.
This research is descriptive-analytical and interpretive based on structuralist approach and empirical evidence obtained from study cases. This approach is used to read public sphere in Bandung's urban spaces to get a reference to the interpretation of relationships from an empirical condition. The degree level and its meaning in Alun-alun Bandung Square with surrounding buildings has been changed and has blocked the accessibility to some buildings and their facades.
Keywords: ***form, function meaning relationship, public sphere, degree level of publicness***

Content from this work may be used under the terms of the Creative Commons Attribution 3.0 licence. Any further distribution of this work must maintain attribution to the author(s) and the title of the work, journal citation and DOI.
Published under licence by IOP Publishing Ltd

1. Introduction

Public spaces in the city of Bandung have been much developed and carried out improvements to infrastructure facilities and the structure of the city, but in the management of the city, private parties are more involved which limits the accessibility to these public places. One barometer of a city's quality of life is the public sphere. Public sphere as a public space that brings together the community of a city's citizens. The public sphere is consider to be a fuser of all interests, citizens, rulers, and the entrepreneurs (1). In the development of life public sphere can include various access areas used together in the forms of roads, parks, shopping, campus, hospitals, public transportation facilities, museums, and so on (2). The development of tourism places in Bandung city and shopping centers (mall), which also increases by providing public space as one element in the design, provides an alternative public sphere in the city of Bandung as a whole. Public sphere in Bandung city region that used to be part of the current political and cultural activities, shifted into a framed of interaction space without any relationships of functions and form around it.

The public sphere as a container of social interaction has changed the meaning and value of space that is more concerned with the interests of individuals and certain groups, and only to gain consumer satisfaction of urban people (3). Public sphere is no longer as a means of communication to shape the process of human social civilization, but only for the needs of urban people who want to seek self-satisfaction. In general, the development of public space, from time to time, has a function as a space of social interaction for the community, the space of people's economic activity and space of cultural appreciation. Public needs in public space always change, in line with the times and the influence of needs, from 4 (four) aspects: 1. Social culture: The human need expresses itself to fellow human beings and to their Lord; 2. Politics: The need for a symbol of city, state, power, government, or people's power; 3. Economy: the need to gather people in a space will encourage the emergence of economic activities, such as markets and so on (4). Bandung as a creative city of the world (UNESCO) provides a political policy on the development of public spaces in several areas in the city of Bandung.

Privatization of public space in Bandung has hampered the public in using the public sphere maximally. Information about public sphere and community activity pattern of Bandung city have not been obtained by city society, but it is related to society need for social interaction, economic activity, and space of cultural appreciation. The most ideal public sphere is the space that everyone can access, without discrimination. Basically, a commercial space is not addressed as a public space because in the commercial space there are imposed conditions that make the space more private. Public sphere co-opted by power are public spaces dominated by user actors; Actors who do not grow from the public but appear in public and occupy the public sphere. Private parties from large-scale economic and business sectors and dominant government organizations that can take over the public sphere formally (5).

In fact, public sphere is not absolutely social or public (open) that anyone can enter into the interaction activities freely. There is a limit to use a public space wherein there are individual and group interaction activities. The degree of perceived publicness in a public sphere can affect how the activities that occur in there, and whether the public space in there, is open for the public or for a particular group (6). Akkar M (2005) states that the degree of publicity in a public space can be identified through three things: Accessibility, Actors, and Activities. Based on these three factors, the degree of publicity (extent of publicness) depends on three indicators such as the following. Accessibility affects how freely the public space can be accessed by the public (community). Actors include the public who are in the space, to what extent are managed and controlled by public actors and used by the public. The last one is the extent to which it can serve the public interest (7). The focus in this study is on the degree of publicness in a different public sphere, depending on the use and ownership of space, whether owned by the government or owned by a particular institution. The degree of publicness of a space also depends on accessibility, who it carries and how far it can serve the public interest. The higher is the mix of different types of individual societies, the higher is the

IOP Conf. Series: Materials Science and Engineering **636** (2019) 012021 doi:10.1088/1757-899X/636/1/012021

degree of publicness. Pattern and form of public space also determine the degree of publicness, the criteria are public space in the building or become part of the city space covered by the building. Markus Zahnd said that the concept of cities in Asia tends to make the city space as a result of the building mass (8). This research will be conducted on the public spaces of Bandung city, especially at Alun-alun Square Bandung, which will identify the form, function, and meaning of public sphere to degree level of publicness. This research is focused on the relationship among function, form, and meaning of public space to the degree level of publicness in Alun-alun Square, Bandung city region.

2. Materials and methods

2.1. Study Case

The study case is Alun-alun Bandung Square in Bandung City, West Java Province, Indonesia. The criteria for the selection of the study case are 1) The object of study represents the public sphere with the degree level of publicness in the space of Bandung city. Study Objects are buildings and open spaces, run by private and government; 2) The object of study represents the development of public sphere in Bandung city based on certain periods. Periodization will show the relationships and values of patterns that have been established and survived from the past until now, although research studies will be focused on current conditions with synchronic approaches.

Figures 1. a.Indonesia Map, b.Bandung City Map, c. Alun-alun Square Map

2.2. Method

The steps taken in this study are as follows: firstly, based on the criteria that have been set in the beginning, the physical boundaries of Alun-alun Bandung city area are determined (9); secondly, the elements that formed the area of Alun-alun Bandung are identified and classified based on the concepts of Nodes, Paths, Landmarks, Districts, and Edges by Kevin Lynch (10); thirdly, recording and identifying the physical condition of the public sphere and the degree level of publicness, as well as acitivities and the surrounding activities, are used to describe the pattern of changes occurring in function and form of the public sphere by architectural ordering principle theory (11), and to relate them to the theoretical elaboration of public space and the degree level of publicness (12). Fourthly, undertaking the study of the meanings that result from the pattern of changes that occur on the meaning of form and function of public sphere to the degree level of publicness; (13). Interpret the results and findings that arised, as well as drawing conclusions obtained through the process of analysis of the relation of form, function and meaning of public space to the degree level of publicness.

3. Results and Discussion

3.1. Determination of Area In Alun-alun Bandung Square

The determination of the physical boundaries refers to the concept of Tipho-Morphology approach, which represents the development of public sphere in Bandung city based on certain periods. Periodization will show the relationships and values of patterns that have been established and survived from the past until now.

Figures 2. Tipho-Morphology Analysis in Alun-alun Bandung Square Area

When it was first built, Alun-Alun Bandung was the front yard of the governmental building as keresiden Bandung in 1889. Then Alun-alun Bandung became an open space used for gatherings and football games. In 1970, Alun-Alun Bandung was a public sphere that could be visited at all times by all people and also it was an access path to The Great Mosque of Bandung through the stairway that fly over *Jalan Dewi Sartika.* In 1980, Alun-Alun became the front yard/garden for The Great Mosque of Bandung and was surrounded by fences (14). Alun-alun Bandung Square was renovated several times until 2014. Then on December 31, 2014, Alun-alun Bandung Square was inaugurated by the Mayor of Bandung, Ridwan Kamil, with many new facilities. The 1,200-square-meter park stands on the roof of a parking lot and is covered with synthetic green grass. The square feature amenities such as a children's playground, library, and WiFi network. At the north of the square, there is a bus stop long enough to complete the facilities in this downtown park.

3.2. Elements of Urban Form That Identify in the Study Case

The elements that formed the area of Alun-alun Bandung are identified and classified based on the concepts of Nodes, Paths, Landmarks, Districts, and Edges by Kevin Lynch (1966).

IOP Conf. Series: Materials Science and Engineering **636** (2019) 012021 doi:10.1088/1757-899X/636/1/012021

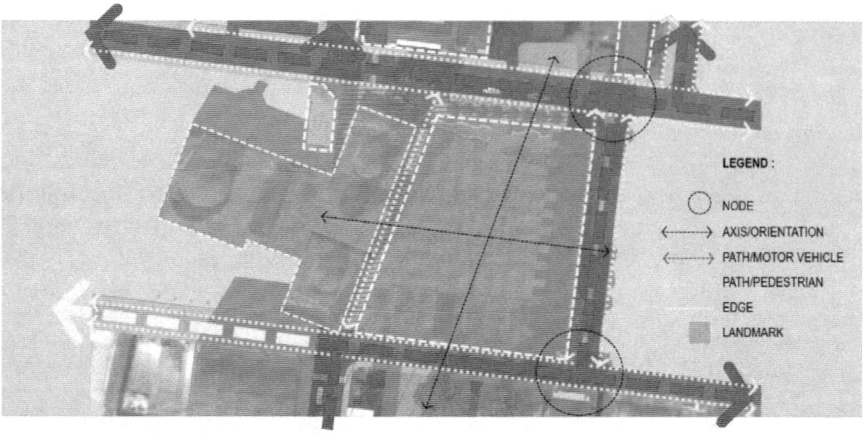

Figures 3. Plan Analysis in Alun-Alun Bandung Square

In the Alun-alun of Bandung city, there are 1 main node and 4 supporting nodes. The main node is the Alun-alun Square, which is the center of regional activities, surrounded by buildings with various functions. Furthermore, the main street in the Alun-alun Bandung square shows a path which connects 4 supporting nodes. Masjid Agung as a landmark element is at the front yard as the district identity.

Figures 4 Degree Level in Alun-Alun Bandung Square

3.3. Physical Condition and Activities in Study Case

Zoning Analysis show the activities that occur therein and whether the public sphere has an open nature for the public or for a particular group.

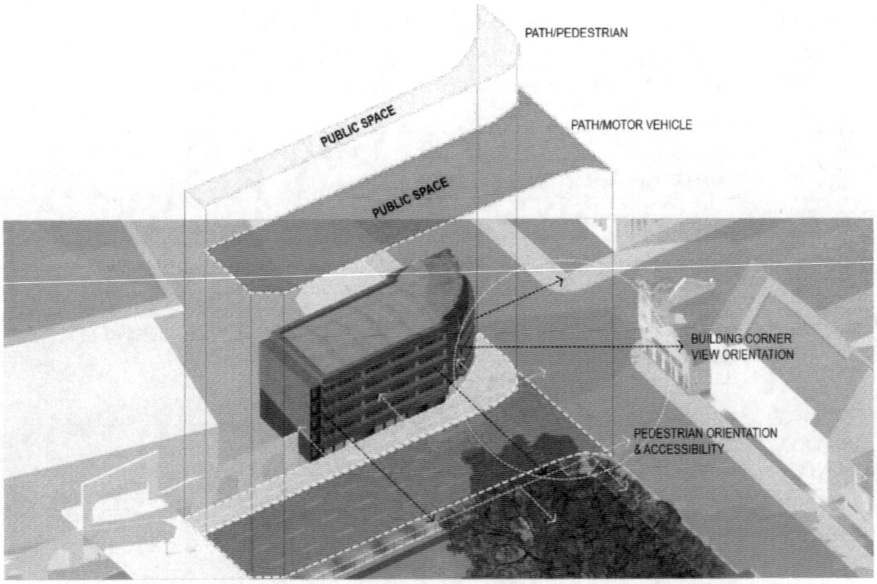

Figures 5. Zoning Analysis in Alun-Alun Bandung Square

Based on the results of the analysis in step two, Alun-alun square was recording and identifying the physical condisitions (Form) and the surrounding activities, to describe the pattern of changes occurring in function and form of the public sphere by architectural ordering principle theory.

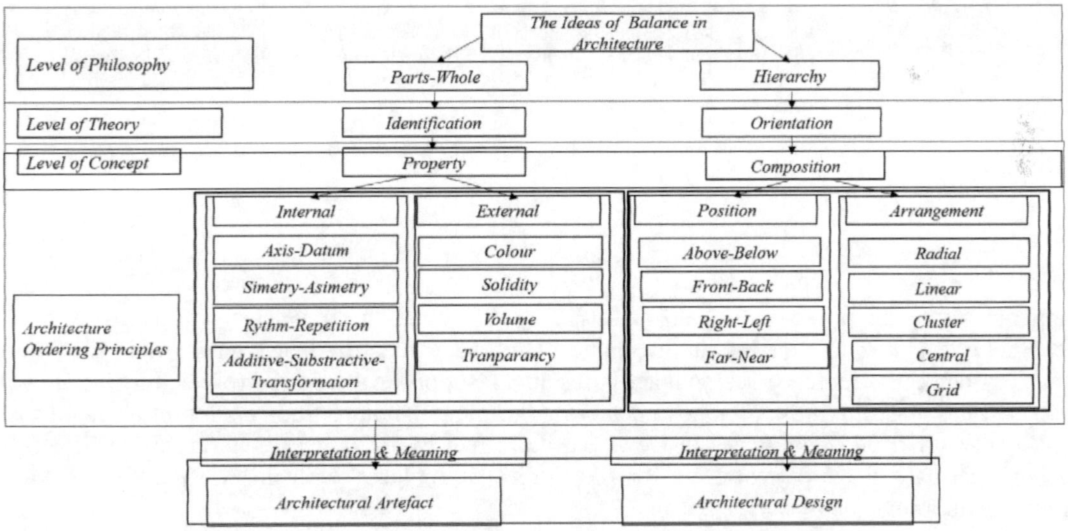

Figures 6. Architectural Ordering Principle, Salura (2018)

Table 1. Change of Function and Form in Alun-alun Bandung Square

ELEMENTS URBAN FORM	**CHANGE IN FUNCTION**		**CHANGE IN FORM**	
	PARTS-WHOLE	**HIERARCHY**	**COMPOSITION**	**PROPERTY**
LANDMARK ELEMENT MASJID AGUNG	Element Functions are the main activities in Alun-alun Bandung Square.	The initial function of the Element has never changed, Element has a function as a generator activity of the area.	The position for the element is on the west side of the Alun-alun square.	Elements with arabic style and the modern buildings are repeated by the surrounding buildings.
MAIN NODE ELEMENT ALUN-ALUN BANDUNG SQUARE	Element Functions are the main activities in Alun-alun Bandung Square	Primary Element Functions as Regional Activity Center, and in the beginning functions as the front yard of the governmental building, and became the front yard/garden for The Great Mosque of Bandung.	Element as the center of Building Orientation .	Primary element has a symmetry shape with the main street and shows a path which connects 4 supporting nodes. Elements are surrounded by buildings with various functions.
PATH-NODE ELEMENT DEWI SARTIKA STREET	Element was part of the Dewi Sartika Street which consisted commercial and office activities. and became a apart of Masjid Raya Building.	Element is part of Masjid Raya Jawa Barat Extention Building, and has no access activity	The street corridor was formed by Path-Node elements which consisted of 2 nodes that had corner buildings, and now the Path-Node element is no longer a corridor/street element.	Corner building as disrict identity, and part of supporting nodes with various buiding styles, Art Decos, and Modern Styles..

From the results of observations and analysis, it was found that changes of Function, Form, and Meanings (15) in Alun-alun Bandung Square have been identified in 1 main node, in Alun-alun Square, which is the center of regional activities, and 4 supporting nodes surrounded by buildings with various styles and activities. Changes occur in Alun-alun Bandung Square supporting node that has significant shift from Main Acces in Dewi Sartika Street to the Access in Asia Africa Street. Street corridor that used to be part of Dewi Sartika Street is replaced with the Masjid Raya Jawa Barat Extention Building.

This significant shift has been changed the orientation of the buidings surrounding of Alun-alun Bandung Square. Front elevation of Swarha building and its main entrance are not accessible as they were used to. Function and form of Alun-alun Square have been significantly changed, the function is not accessible to every side of the square, the form is changed from the single square that clearly boundary with the main streets, to the attached square, that have blocked acces and façade of the buildings.

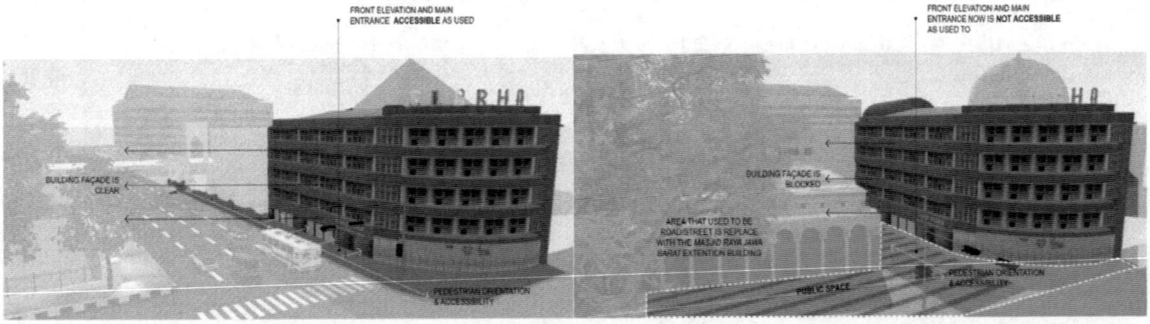

Figures 7. Accessible Analysis (Dewi Sartika Street) in Alun-Alun Bandung Square

3.4. Degree Level of Publicness in the Study Case

The degree level of public space can affect the way the activities occure therein and whether the public sphere has an open nature for the public or for a particular group. Related to this, Sorkin (1992 in Carmona, 2003, p.111) distinguishes public space in three categories: 1. External public space, is accessible to all and is a real sense of public space. Usually a land located in a private property. Such as square, street, park, parking area, beach, forest, lake, etc; 2. Internal public space, have more limited access. Examples can be the public institutions such as libraries, museums, town hall, and others include public transport facilities such as trains, bus terminals, airports, etc; 3. External and Internal quasi-public space, is a public space with limited access, in the sense that even if it is 'public', the owner or manager of the space has the right to manage the access time and the user's behavior within. The space is open and there are physical and psychological constraints and this public space is owned by non-governmental parties, ie. private or specific communities. And the example can be restaurants, cinemas, or malls, etc.

Table 2. Degree Level of Publicness and Meanings in Alun-alun Bandung Square

ELEMENTS URBAN FORM	DEGREE LEVEL OF PUBLICNESS	CHANGE OF MEANINGS IN ALUN-ALUN BANDUNG SQUARE
LANDMARK ELEMENT MASJID RAYA BANDUNG	**External and Internal quasi-public space**, Element has a function as worship activities, a public space with limited access, in the sense that even if it is 'public', the owner or manager of the space has the right to manage access time in several private zones, and the user's behavior within.	Domination in Function = Domination in Form ► Meaning of Social & Cultural Ideology
MAIN NODE ELEMENT ALUN-ALUN BANDUNG SQUARE	**External Public Space,** Element is accessible to all, thought the function is not accessible from every side of the square, the form is changed from the single square that clearly has boundaries with the main streets, to the attached square, that have blocked acces and façade of the buildings.	Domination in Function > Domination in Form ► Meaning of Social & Cultural Activity

IOP Conf. Series: Materials Science and Engineering **636** (2019) 012021 doi:10.1088/1757-899X/636/1/012021

PATH-NODE ELEMENT DEWI SARTIKA STREET	**Internal Public Space,** Element has significant shift from Main Acces in Dewi Sartika Street to the Access in Asia Africa Street. Street corridor that used to be part of Dewi Sartika Street is replaced with the Masjid Raya Jawa Barat Extention. Former Building. Corner building as disrict identity with various buiding style, Art deco and Modern style.	Domination in Function < Domination in Form ► Meaning of Physical Building.

there is a hierarchy of meaning from Simbolyc Ideology, Social and Cultural Activity to Physical condition meaning (16). These meanings are related to the degree level of publicness of each conditions in the study case. Alun-alun Bandung Square as the main node with the meaning of Social and Cultural Ideology, has degree level category as External Public Space, that has full access, eventhough there is a change in accessing the surroundings of Alun-alun Bandung Square that blocked the accessibility to some buildings and their facades.

4. Conclusion

This study concluded that firsly, the degree level and its meaning in the public sphere in the Bandung City, Alun-Alun Bandung square's meaning has shifted from the political ideology to the social-cultural ideology meaning. The meaning of shared spaces is shifted from the political-social value frame and the joint gathering function celebrates togetherness into only a field-framed interactions spaces. Secondly, the hierarchy of meaning, from Simbolyc Ideology, Social and Cultural Ideology to Physical condition meaning are related to the degree level of publicness of each conditions of public spaces in the study case. Thirdly, the results of the study confirm that the method used in the study to interpret the meaning of degree level of public space can be used as an alternative for reading the accessibility in urban city areas.

References

[1] Habermas, J. 1974. The public sphere: An Encyclopedia article (1964). New German Critique, 3, 49–55.

[2] Siregar, Sandi Aminuddin ,1990. *Bandung—The Architecture of a City in Development: Urban Analysis of a Regional Capital as a Contribution to the Present Debate on Indonesian Urbanity and Architectural Identity*. Dissertation. Katholieke Universiteit Leuven, Leuven.

[3] Benhabib, S. 1992. *Models of public space: Hannah Arendt, the liberal tradition, and Jürgen Habermas.* In C. J. Calhoun (Ed.), Habermas and the public sphere (Studies in contemporary German social thought, pp. 73–98). Cambridge, MA: MIT Press.

[4] Dick, H.W. and Rimmer P.J., 1997, *Beyond The Third World City: The New Urban Geography Of South-East Asia.* Urban Studies.

[5] Lefebre, Henri, 1991, *The Production Of Space*, Wiley

[6] Habermas, J. 1989. *The structural transformation of the public sphere: An inquiry into a category of bourgeois society* (Studies in contemporary German social thought). Cambridge, MA: MIT Press.

[7] Akkar, M. 2005. *Questioning 'Publicness" of Public Spaces in Post Industrial Cities*, Traditional Dwellings and Settlements Review, p.24.

[8] Zahnd, Markus, 2006, *Perancangan kota secara terpadu : teori perancangan kota dan penerapannya*, Kanisius.

[9] Wibisono, Bambang Hari, 2001, *Transformation of Jalan Malioboro, Yogyakarta: The Morphology and Dynamics of a Javanese Street*. Dissertation. Faculty of Architecture, Building and Planning, The University of Melbourne.
[10] Lynch, Kevin. 1966. *The Image Of The City*, The MIT Press
[11] Salura, Purnama, 2018, *The Philosophy Of Architectural Ordering Principles*. International Journal of Engineering and Technology, Vol7, No.3, p.218-221.
[12] Carr, Stephen, 1995, *Public Spaces*, John Wiley & Sons.
[13] Norberg-Schulz, Christian, 1988, *Architecture : Meaning and Place*, Rizzoli Int. Publication, New York.
[14] Voskuil, Robert P.G.A., 2017, *Bandung, Citra Sebuah Kota*, Penerbit ITB
[15] Salura, Purnama, Fauzi, Bachtiar, 2012, *The Ever-rotating Aspects of Function-Form-Meaning in Architecture*. Journal of Basic and Applied, 7086-7090.
[16] Hanafiah, Ully Irma Maulina; Sudikno, Antariksa; Salura, Purnama. 2018. *Changes in the Meaning of Primary Elements of the European Region in Medan City, Sumatra Island, Indonesia.* International Journal of Engineering & Technology, *v. 7*(n. 4.7), 453–457,. https://doi.org/http://dx.doi.org/10.14419/ijet.v7i4.7.27358.

IOP Conf. Series: Materials Science and Engineering **636** (2019) 012022 doi:10.1088/1757-899X/636/1/012022

Understanding Social Sustainability Criteria for Public School Building

MA Rohman[1] and YA Hidayat[2]
[1,2]Department of Civil Engineering, Institut Teknologi Sepuluh Nopember
Kampus ITS Sukolilo, Surabaya, Indonesia

E-mail: arif@ce.its.ac.id or arif.its@gmail.com

Abstract. The school building has an important role in the education process of the young generation. Surabaya as the second biggest city in Indonesia has a quite large number of school buildings. As a building is a construction project, it has a long project life cycle, it is very important to adopt a sustainable development concept to provide maximum benefit to the community. Social sustainability is one of the sustainable development aspects and has a close relationship with the community needs and interests. This aspect is essential to promote project benefit streams to the community not only in the short-term but also in the long-term. This research is an initial research step to investigate social sustainability criteria in public school buildings in Surabaya. Data was collected through a series of research steps consisting of a preliminary survey, pilot test, and questionnaire distribution. Mean analysis was used to rank the attributes criteria according to the three stakeholders' perceptions. Finally, seventeen attributes were considered relevant to this research. This study provides a reference for the designer as well as the decision-maker such as the government on how to achieve social sustainable public buildings.

1. Introduction

Public school buildings are important facilities for the community, especially for the next young generation. As the second biggest city in Indonesia, Surabaya has a very large number of public buildings. According to [1], 168,578 public school buildings are available ranging from kindergarten to high school / vocational level in 2016/2017. However, in addition to the positive impact, the existence of a public building can also have a negative impact to the community if they are not managed properly during the construction phase or operation phase. It happens in the form of environmental pollution, flooding, traffic congestion, damage to surrounding buildings and others. Therefore, it is very important to ensure the school building project is successful by presenting maximum positive benefits and reducing negative impacts on the community.

From the community perspective, as the building has a long project life cycle, the concept of sustainable development (SD) is quite relevant to be applied in this type of projects [2]. This SD concept stressed that the development must be able to meet current needs without sacrificing the interests of future generations by balancing economic, environment, and social aspects or known as the "triple bottom line" [3]. Unfortunately, while the economic and environmental aspects have received a lot of attention, social sustainability (social aspects) has not been widely explored [4]. Indeed, social sustainability concept is very important to be realized in a project so that it can provide long-term project benefits to all communities at large [5]. Several types of research have been

Content from this work may be used under the terms of the Creative Commons Attribution 3.0 licence. Any further distribution of this work must maintain attribution to the author(s) and the title of the work, journal citation and DOI.
Published under licence by IOP Publishing Ltd

conducted to investigate social sustainability application in the built environment [4, 5, 6, 7]. As limited research has been conducted to investigate social sustainability in public school buildings, this research aims to bridge the gap in the literature by investigating social sustainability criteria in this field.

2. Literature Review

2.1. Sustainability Concept

Literally, Sustainable Development (SD) can be articulated from two words, namely sustainable and development. While "Sustainable" means "can be continued", "Development" can have the meaning the way to increase the community's well-being over time". Therefore, sustainable development can be defined as meeting the community well-being in very long terms. Meanwhile, [3] defines sustainable development as "a development which meets the present's needs and aspirations without comprising the future generations ability to meet their needs". The concept of sustainability stressed to balance the implementation of the three sustainability components, namely economic, environmental, and social aspects.

With regards to the construction sector, [8] argued that sustainable construction should be applied in the overall construction stages since material extraction, planning, design, and building until the demolition which is seen as a holistic process to maintain the harmony between the natural and built environments to present human dignity and promoting economic equity. The implementation of the sustainability concept in a construction project should also balance its three dimensions. However, while the economic and environmental aspects have obtained a lot of attention, social sustainability still has not been widely explored in the construction field. Indeed, this social aspect is also important as the other two components to ensure that projects can deliver their benefits to all stakeholders, including the community in the long term [5].

2.2. Social Sustainability

The concept of social sustainability is actually still in the development process [7]. Even though the social sustainability term has been acknowledged, the definition of this concept is still fuzzy and has not been agreed upon by researchers [4]. According to [9], social is defined as "something relating to society or its organization". Several issues related to social are equity, health, education, housing, security (combating crime), population/demographics, poverty, and governance [10]. By referring to the definition above, literally social sustainability means "the ability to maintain the society and something related to it at a certain rate or level".

In this paper, social sustainability is defined as "a life-enhancing condition within communities, and a process within communities that can achieve that condition" [11]. It consists of several principles such as equity, diversity, interconnectedness, quality of life, and democracy, and governance. Several social sustainability criteria from previous literature were identified as in Table 1. Several studies have also been conducted in the area of the built environment [5, 6, 7, 12]. Yet, how this social sustainability can be applied in the public school building has not been widely explored.

3. Research Method

This research adopts a quantitative approach using a statistical approach based on respondents' perceptions. The data was acquired through a series of data collection steps, namely preliminary survey, pilot test, and main survey. A preliminary survey was conducted to verify research variables with the study context. Four experts were interviewed regarding the research attributes. Following the main survey, the analysis was conducted to obtain attributes which were considered relevant according to the expert opinions. According to the experts, it was found that all 17 variables were considered relevant for this research.

A pilot test was then conducted before the main survey through questionnaire distribution to ensure that the prepared questionnaire can be well understood by the respondents. Respondents involved in this study can be categorized into three sectors, namely user, neighborhood and construction sector. This classification adopts [6] stakeholder's classification. User consists of the student, teacher and school employ while the surrounding community is people who live in the school surrounding area. Meanwhile, the construction sector includes a consultant, contractor, and supplier. There is no specific rule according to the samples required for the statistical analysis. However, this research sought to get a large number of samples, as according to the statistic test, the larger data can be obtained, the better it can represent the population. Data was analysis using a mean as an appropriate statistical technique [13] to obtain attributes which are considered important for the school building.

4. Result and Discussion

4.1. Respondent Background

From the main survey, finally, 100 samples were received from three stakeholders, namely user, construction provider, and the community. Table 1 presents the respondent's background and characteristic. According to Table 1, 30% of respondents in this study had senior high school backgrounds, while 70% held status as a bachelor. Meanwhile, respondents with age between 31-40 years are the largest component of the respondents. Finally, most of the respondents are holding minimal bachelor degree with 70% composition.

Table 1. Social Sustainability Criteria

No.	Criteria	Source
1	Building can provide security for all people.	[4, 6, 12]
2	Location of the building is close to public transportation facilities.	[12]
3	Building provides an open space area.	[6, 12]
4	Building can be accessed by all people without any gender or social status restrictions.	[12, 14]
5	The building can promote the local community's needs.	[4, 12, 14]
6	Building can adopt to changes in the surrounding community.	[12]
7	The level of pollution (water, air, waste) from the building existence can be tolerated.	[4, 12, 14]
8	The surrounding community is involved in the decision-making process during the construction and operation phases.	[4, 12]
9	Building becomes facility for training and education.	[4, 11, 15]
10	Building existence contributes to maintain the local community value and culture.	[11]
11	Building existence can be identity and historical value for the surrounding community.	[11]
12	The building can create a sense of belonging and pride for users and the surrounding community	[4, 7, 14]
13	Building existence is supported by users and the surrounding community.	[7, 14]
14	Building existence can improve the economy of surrounding community	[4, 6, 7]
15	The existence of buildings can create new jobs for the surrounding community	[4, 6, 16]
16	Building existence does not damage the community social order.	[4, 15]
17	The expected outcomes of the building construction have been communicated to the community	[16]

4.2. Reliability Test

Reliability test is a series of instrument measurement tests to ensure that the measurement has consistency if it is used repeatedly [13] which is usually calculated using the Cronbach Alpha coefficient. To be reliable, this coefficient value has to be greater than 0.60 [12]. According to the analysis the reliability of the questionnaire was 0.671, which means this survey instrument's reliability can be considered sufficient.

4.3. Attribute Ranking Based on Stakeholder Perceptions

Attribute mean analysis aims at understanding the importance of the attributes according to the respondents' perception, which is shown in Table 2. The ranking is presented according to overall stakeholder, user, neighborhood, and construction sectors. The first ranking attribute from the overall stakeholder is that The building is secure for all people with the mean 4.64. Indeed, security is a basic need that should be fulfilled in every facility provided for the community [4]. This means that every building should become a place, which can prevent a criminal action for the user and surrounding community. However, while neighborhood sector also ranked these attributes as the first ranking attribute, the user and construction sectors placed these attributes in the second and third positions, respectively.

Table 2. Repondents background

Group Category	Group Type	Percent (%)	Percent (Cumulative)
Stakeholder Type	Construction Type	33	33
	User	35	68
	Neighborhood Community	32	100
Age	10-20 years	13	13
	21-30 years	25	38
	31-40 years	36	74
	41-50 years	22	96
	51-60 years	4	100
Education Category	Junior High School	0	0
	Senior High School	30	30
	Bachelor Degree and Above	70	100

The second important attribute from the overall perspective is The building location is close to transportation facility (mean 4.55). Almost similar to that, the user, neighborhood, and construction sectors also placed this attribute in the second and third rankings. This result shows that transportation is very important to the community as this facility can provide accessibility to the building as an important indicator of social sustainability [12]. Meanwhile, the attribute of Building becomes a facility for knowledge and skill development (mean 4.52) which is ranked as the third position according to the overall sector. The building is a strategic facility, which provides an opportunity for the students or community to develop their knowledge and skills [5] either during its construction or operation phases. However, the three stakeholders had different perceptions of this attribute. While the user put this attribute as the main priority or the first ranking, the neighborhood and construction sectors placed this attribute as the tenth and fifth positions, respectively.

The fourth attribute is The building is able to improve the economy of the surrounding community (4.37). This attribute is perceived as important because a public school usually attracts many people to come and meet that place. As the people gather in one place, it is possible to stimulate business activities in its surrounding area [7]. Yet, similar to the previous attribute, the three stakeholders also had a different perspective for this attribute. If the construction sector positioned this attribute as the first ranking, the user and neighborhood ranked this attribute in the fourteenth and fifth positions, respectively. Finally, the fifth position attribute from the overall sector is The building which provides an open space area with mean value 4.36. In order for the people to be able to do their activities productively, a sufficient open space is needed. Human needs sufficient open space as a facility for relaxing and putting the stress away [12]. Even, for a public school building, open space is really needed to allow students explore their creativity as well as developing their skills through the surrounding environment. An open space is needed to make the user easy to use the facility according to their approach and use [17]. This is also important for the surrounding community where there should be sufficient open space between the school building and their settlement area. This is important to avoid some negative effects from the building, such as noise, traffic congestion and some other similar effects. Unfortunately, the three stakeholders also had a different point of view regarding the attribute ranking. While the construction sector ranked this attribute as the fourth position, the user and the community placed this attribute in eighth and ninth positions, respectively.

IOP Conf. Series: Materials Science and Engineering **636** (2019) 012022 doi:10.1088/1757-899X/636/1/012022

Table 3. Mean Analysis and Attributes Ranking

No.	Criteria	All Stakeholders		User		Neighborhood		Construction Sector	
		Mean	Ran-king	Mean	Ran-king	Mean	Ran-king	Mean	Ran-king
1	Building can provide security for all people.	4.64	1	4.73	2	4.66	1	4.55	3
2	Location of the building is close to public transportation facilities.	4.55	2	4.57	3	4.50	2	4.58	2
9	Building becomes facility for training and education.	4.52	3	4.74	1	4.31	10	4.48	5
14	Building existence can improve the economy of surrounding community	4.37	4	3.97	14	4.44	5	4.64	1
3	Building provides an open space area.	4.36	5	4.20	8	4.34	9	4.55	4
4	Building can be accessed by all people without any gender or social status restrictions.	4.36	6	4.31	6	4.41	6	4.36	8
10	Building existence contributes to maintain the local community value and culture.	4.30	7	4.20	9	4.36	8	4.35	9
13	Building existence is supported by users and the surrounding community.	4.25	8	4.31	7	4.14	14	4.44	6
15	The existence of buildings can create new jobs for the surrounding community	4.25	9	4.40	4	4.38	7	3.97	13
11	Building existence can be identity and historical value for the surrounding community.	4.19	10	3.97	14	4.22	11	4.39	7
17	The expected outcomes of the building construction have been communicated to the community	4.17	11	3.86	17	4.53	3	4.15	10
16	Building existence does not damage the community social order.	4.16	12	3.89	16	4.63	2	4.00	12
7	The level of pollution (water, air, waste) from the building existence can be tolerated.	4.10	13	4.37	5	4.19	13	3.73	17
12	The building can create a sense of belonging and pride for users and the surrounding community	4.06	14	4.09	10	4.05	15	4.06	11
6	Building can adopt to changes in the surrounding community.	4.01	15	4.06	12	4.19	12	3.79	16
5	The building can promote the local community's needs.	3.95	16	4.06	11	3.87	17	3.94	14
8	The surrounding community is involved in the decision-making process during the construction and operation phases.	3.93	17	3.91	15	4.03	16	3.85	15

The results show that three stakeholders had different perceptions to rank the five highest attributes' positions. The agreement is important in order for it to be used as the guidance for the designer or the decision-maker such as the government to plan and design the building appropriately and to ensure that social sustainability can stream to the community to maximize the project benefit as well as minimizing its negative impact.

5. Conclusion

This study found 17 attributes to measure social sustainability in public school building in Surabaya. This social sustainability attribute is very important to promote a holistic concept of sustainable development by balancing three sustainability components, namely economic, environmental and social aspect. Accommodating social aspect in the success criteria which is close to the community needs and interest can be expected to improve the stakeholder's support to the project and to minimize the stakeholders' opposition to achieve a more successful public school building project. It was also found that the stakeholders still had a disagreement regarding the attributes' priority. This means there is a need to develop communication between the stakeholders through a gathering moment, such as public participation, to better develop stakeholders' understanding regarding the important social sustainability criteria for this type of project.

Acknowledgement
The authors would like to thank the Ministry of Research, Technology and Higher Education who provided a grant for this research in 2019 as well as several experts and respondents who have been involved in the data collection processes, both the preliminary survey as well as the main survey.

References
[1] The Ministry of Education and Culture (Kementrian Pendidikan dan Kebudayaan). (2017). Ikhtisar Data Pendidikan tahun 2016/2017.
[2] Kumaraswamy, M.M., Ling, F.Y.Y., Anvuur, A.M. and Rahman, M.M. (2007). *"Targeting relationally integrated teams for sustainable PPP's"*. Engineering, Construction and Architectural Management, Vol. 14 No. 6, pp. 581-595.
[3] World Commission on Environment and Development. (1987). Our common future. New York.
[4] Colantonio, A., Dixon, T., Ganser, R., Carpenter, J. And Ngombe, A. (2009). Measuring socially sustainable urban regeneration in Europe. Oxford Institute for Sustainable Development (OISD). School of the Built Environment, Oxford Brookes University.
[5] Doloi, H. (2012). *"Assessing stakeholders' influence on social performance of infrastructure projects"*. Facilities, vol. 30 Nos 11/12, pp. 531-550.
[6] Almahmoud, E.S. and Doloi, H. (2015). *"Assessment of social sustainability in construction projects using social networks analysis"*. Facilities, Vol. 33 Nos 3/4, pp. 152-176.
[7] Rohman, M.A., Doloi, H., and Heywood, C, A. (2017). *"Success criteria of toll road projects from a community societal perspective"*. Built environment Project and Asset Management, Vol. 7 Iss 1 pp. 32-44.
[8] Du Plessis, C. (2002). Agenda 21 for sustainable construction in developing countries. A discussion document. The International Council for Research and Innovation in Building and Construction CIB and United Nations Environment Programme International Environmental Technology Centre UNEP-IETC.
[9] Oxford Dictionary. (2016). Available at: http://oxforddictionaries.com/.
[10] United Nation Department of Economic and Social Affairs. (2007). Indicators of Sustainable Development: Guidelines and Methodologies. Third Edition. United nations, New York.
[11] McKenzie, S. (2004) Social sustainability: towards some definitions. Hawke Research Institute, University of South Australia.
[12] Chan, E. and Lee, G.K.L. (2008). *"Critical factors for improving social sustainability of urban renewal projects"*. Social Indicator Research, No. 85 No. 2, pp. 243-256.
[13] Vogt, W.P. (2007). Quantitative Methods for Professionals, Boston: Pearson.
[14] Commission for Architecture and the Built Environment (CABE) and Department of the Environment, Transport and the Regions (DETR). The Value of Urban Design, Bartlett School of Planning commissioned by Commission for Architecture and the Built Environment, London.
[15] Littig, B. and Grießler, E. (2005). *"Social sustainability: a catchword between political pragmatism and social theory"*. International Journal of Sustainable Development, Vol. 8 Nos 1/2, pp. 65-79.
[16] Rodolfo, V.V. (2011). Social sustainability consideration during planning and design: A framework of Processes for construction projects. Clemson University.
[17] Rohman, M.A and Wiguna, I.P.A (2019), *"Evaluation of road design performance in delivering community project social benefits in Indonesian PPP"*, International Journal of Construction Management. DOI : 10.1080/15623599.2019.1603095.

Institute of Physics
Dirac House, Temple Back
Bristol BS1 6BE UK

ISSN: 1757-8981
ISBN 978-1-5108-9838-7

CURRAN ASSOCIATES INC.
proceedings.com